From Complexity to Life

FROM COMPLEXITY TO LIFE

*On the Emergence of
Life and Meaning*

Edited by
Niels Henrik Gregersen

OXFORD
UNIVERSITY PRESS

2003

OXFORD
UNIVERSITY PRESS

Oxford   New York

Auckland   Bangkok   Buenos Aires   Cape Town   Chennai
Dar es Salaam   Delhi   Hong Kong   Istanbul   Karachi   Kolkata
Kuala Lumpur   Madrid   Melbourne   Mexico City   Mumbai   Nairobi
São Paolo   Shanghai   Taipei   Tokyo   Toronto

Copyright © 2003 by Oxford University Press, Inc.

Published by Oxford University Press, Inc.
198 Madison Avenue, New York, New York 10016

www.oup.com

Oxford is a registered trademark of Oxford University Press

Library of Congress Cataloging-in-Publication Data

From complexity to life : on the emergence of life and meaning / edited by Niels
Henrik Gregersen.
p. cm.
Includes bibliographical references and index.
ISBN 0-19-515070-8
1. Religion and science—Congresses.   2. Complexity (Philosophy)—Congresses.
I. Gregersen, Niels Henrik, 1956–
BL240.3 .F76 2002
501—dc21   2001052389

2 4 6 8 9 7 5 3 1

Printed in the United States of America
on acid free paper

ACKNOWLEDGMENTS

This volume of essays came out of a research symposium, "Complexity, Information, and Design: A Critical Appraisal," held in Santa Fe, New Mexico, on October 14–16, 1999. Led by the physicist Paul Davies, 10 mathematicians, physicists, theoretical biologists, and theologians met under the aegis of the John Templeton Foundation in order to explore the broader issues raised by information theory and complexity studies, in particular their impact on our worldview and sense of meaning.

We wish to thank the John Templeton Foundation for its sustained interest in the field of complexity research, an interest also marked by the presence of Sir John Templeton and the executive director of the Foundation, Dr. Charles L. Harper, at the Santa Fe symposium. We owe a special thanks to senior fellow Dr. Mary Ann Meyers, who coorganized the symposium and helped the editorial process with her characteristic blend of efficiency and humor.

While acknowledging that we are only beginning to understand, we hope that we have been able to summarize some key elements in recent research and to provide a road map for future investigations into metaphysical questions related to complexity studies.

Chapter 2, Gregory J. Chaitin, "Randomness and Mathematical Proof," originally apeared in *Scientific American* 232 (May 1975): 47–52. Chapter 3, Charles Bennett, "How to Define Complexity in Physics and Why," originally appeared in *SFI Studies in the Sciences of Complexity*, vol. 8, ed. W. H. Zurek (Reading, Mass.: Addison Wesley, 1990):

137–148. Chapter 4, Stuart Kauffman, "The Emergence of Autonomous Agents," originally appeared as chapter 1 in Kauffman, *Investigations* (New York: Oxford University Press, 2001), and has been slightly abbreviated and modified.

Aarhus, Denmark
June 16, 2001   N. H. G.

CONTENTS

CONTRIBUTORS

CHARLES H. BENNET is senior scientist at IBM's Thomas J. Watson Research Center in Yorktown Heights, New York. He coinvented quantum cryptography and is a codiscoverer of the quantum teleportation effect. In his many research papers he has done groundbreaking work on the definition of the "logical depth" of complex objects.

GREGORY J. CHAITIN is at IBM's Thomas J. Watson Research Center in Yorktown Heights, New York, and is also an honorary professor at the University of Buenos Aires and a visiting professor at the University of Auckland. He is a principal architect of algorithmic information theory, and the discoverer of the celebrated Omega number. His books include *Algorithmic Information Theory* (1987), *The Limits of Mathematics* (1998), *The Unknowable* (1999), *Exploring Randomness* (2001), *and Conversations with a Mathematician* (2002).

PAUL DAVIES has held chairs at the University of Newcastle-upon-Tyne, England, and at the University of Adelaide, Australia, and is currently visiting professor at Imperial College, London. Apart from his scientific work in the field of quantum gravity and cosmology, he has written many bestselling books on the philosophy of nature, especially physics. In his most recent book, *The Fifth Dimension: The Search for the Origin of Life* (1998), he examines the source of biological information.

WILLIAM A. DEMBSKI is associate professor in conceptual foundations of science at Baylor University, Texas, and a fellow at the Center for the Renewal of Science and Culture at the Discovery Institute in Seattle. He is author of *The Design Inference: Eliminating Chance Through Small Probabilities* (1998) and *Intelligent Design: The Bridge between Science and Theology* (1999).

NIELS HENRIK GREGERSEN is a research professor in theology and science at the University of Aarhus, Denmark, and President of the Learned Society in Denmark. He is editor of *Studies in Science and Theology*, and his recent books include *Rethinking Theology and Science* (1998), *The Human Person in Science and Theology* (2000), and *Design and Disorder* (2002).

STUART A. KAUFFMAN is a founding member of the Santa Fe Institute, New Mexico, and a founding general partner of the BiosGroup LP. His pioneering work in complexity studies include numerous research papers and the highly influential books *The Origins of Order: Self-Organization and Selection in Evolution* (1993), *At Home in the Universe* (1996), and *Investigations* (2000).

WERNER R. LOEWENSTEIN is the director of the Laboratory of Cell Communication in Woods Hole, Massachusetts, and a professor and chairman emeritus of physiology and biophysics at the University of Miami, and former professor and director of the Cell Physics Laboratory, Columbia University, New York. He is editor-in-chief of the *Journal of Membrane Biology*, and among his many other publications is *The Touchstone of Life: Molecular Information, Cell Communication and the Foundations of Life* (1999).

HAROLD J. MOROWITZ is the Clarence Robinson Professor of Biology and Natural Philosophy at George Mason University and professor emeritus of molecular biophysics and biochemistry at Yale University. He was editor-in-chief of the journal *Complexity* from 1995 to 2002 and, apart from many books, including *Beginnings of Cellular Life: Metabolism Recapitulates Biogenesis* (1992) and *The Kindly Dr. Guillotine* (1997), he has contributed widely to the field of complexity studies.

ARTHUR PEACOCKE worked in the field of physical biochemistry for more than 25 years. He was the founding director of the Ian Ramsey Centre at Oxford University, which he directed until recently. Apart from a large scientific production, including *An Introduction to the Physical Chemistry of Biological Organization* (1989), his works include a series of world-renowned books on the relation between science and religion, including *Theology for a Scientific Age* (1993) and *Paths from Science Towards God* (2001).

IAN STEWART is director of the Mathematics Awareness Centre at Warwick University, professor of mathematics at Warwick, and a fellow of the Royal Society. He is a columnist for *Scientific American* and the author of some 60 books in mathematics and physics, including the highly acclaimed *Nature's Numbers: The Unreal Reality of Mathematics* (1996), *Life's Other Secret: The New Mathematics of the Living World* (1998), and *The Science of Discworld* (1999). He is also a writer of science fiction.

From Complexity to Life

ONE

INTRODUCTION: TOWARD
AN EMERGENTIST WORLDVIEW

Paul Davies

It takes only a casual observation of the physical universe to reveal that it is awesomely complex. On an everyday scale of size
we see clouds and rocks, snowflakes and whirlpools, trees and people
and marvel at the intricacies of their structure and behavior. Shrinking
the scale of size we encounter the living cell, with its elaborate cus-
tomized molecules, many containing thousands of atoms arranged with
precise specificity. Extending our compass to the cosmos, we find complexity on all scales, from the delicate filigree patterns on the surface of
Jupiter to the organized majesty of spiral galaxies.

Where did all this complexity come from? The universe burst into
existence, possibly from nothing, in a big bang about 13 billion years
ago. Astronomers have discovered strong evidence that just after the
big bang the universe consisted of a uniform soup of subatomic particles
bathed in intense heat radiation. Some theories suggest that this state
may have been preceded, a split second after the cosmic birth, by little
more than expanding empty space. In other words, the universe started
out in a simple, indeed almost totally featureless, state. The complexity,
diversity, and richness that we see today on all scales of size have emerged
since the beginning in a long sequence of physical processes. Thus the
complexity of the cosmos was not imprinted on it at the beginning but
was somehow implicit in the laws of nature and the initial conditions.

Scientists would like to understand both the nature and origin of this
complexity. At first sight, however, the world seems so hopelessly complicated as to lie beyond human comprehension. The audacious asser-

3

tion of the scientific enterprise is that beneath this surface complexity lies a deep and unifying simplicity. Physicists expect that the underlying laws of the universe are elegant and encompassing mathematical principles. They dream of a final theory in which all the basic laws are amalgamated into a single overarching mathematical scheme, perhaps even a formula so compact and simple that you might wear it on a T-shirt.

How does the complexity of the world flow from the operation of simple laws? The atomists of ancient Greece hypothesized that the cosmos consists entirely of indestructible particles (atoms) moving in a void, so that all physical structures, however complex, are but differing arrangements of atoms, and everything that happens in the physical world is ultimately just the rearrangement of atoms. And that is all. Although the objects we now call atoms are not the fundamental particles of the Greeks but composite bodies that can be broken apart, the essential idea lives on. At some sufficiently small scale of size, physicists believe, there is a set of primitive, indecomposable entities (perhaps not particles but strings, membranes, or some more abstract objects), in combinations of which the familiar world of matter and forces is constructed. It is in this bottom level of structure that the unified theory would operate, weaving simple building blocks together with simple interactions. Everything else—from people to stars—would stem from this microscopic frolic.

Reductionism and Beyond

The philosophy that the whole is nothing but the sum of its parts is known as reductionism, and it has exercised a powerful grip on scientific thinking. If all physical systems are ultimately nothing but combinations of atoms, or strings, or whatever, the world is already explained at the level of subatomic physics. The notion of complexity then loses most of its significance, since the real explanatory power of science lies at the lowest level of description. All the rest is qualitative and woolly embellishment. Thus the claim, for example, that a human being is "alive" is robbed of its force, because no atom in the human body is living, and human bodies are nothing but combinations of atoms. The quality of "being alive" is effectively defined away. Similarly, if brains are nothing but collections of cells that are in turn collections of atoms, then the mental realm of thoughts, feelings, sensations, and so on is equally vacuous.

An out-and-out reductionist has no interest in complexity, since a complex macroscopic system differs from a simple one only in the specific arrangements of its parts, and if you suppose that the arrows of explanation always point downward—to the bottom level—then if the bottom level is explained, so is the whole. That is why unified final theories are sometimes called "theories of everything."

Few scientists take this extreme position, of course. Most are pragmatists; they acknowledge that even if in principle subatomic physics could explain, say, the operation of the human brain, in practice there is no hope of doing this. Some go even further and assert that complex qualities like "being alive, "thinking," and so on have genuine significance that cannot, *even in principle*, be reduced to the behavior of subatomic particles. At each level of description, from atoms, up through molecules, cells, organisms, and society and culture, genuinely new phenomena and properties emerge that require laws and principles appropriate to those levels. These "emergenticists" refute out-and-out reductionism as "the fallacy of nothing-buttery," which fails to explain qualities like "being alive" but instead merely defines them away. In his chapter Harold Morowitz itemizes the emergent conceptual levels and discusses how nonreducible laws and principles might operate within them.

Once it is accepted that "being complex" is a genuine physical attribute that deserves an explanation, the problem immediately arises of how to characterize and quantify complexity. Until recently complexity was regarded as mere complication. Today, however, researchers recognize that complex systems display systematic properties calling for novel mathematical descriptions. The establishment of research centers, such as the Santa Fe Institute for the Study of Complexity, devoted to uncovering quasi-universal principles of complexity has led to an enormous advance in our understanding of the subject. The availability of fast computer modeling has enabled complex systems to be studied in great generality. It holds out the promise that systems as disparate as economies, immune systems, and ecosystems might conform to common mathematical principles. Complexity theory is also illuminating the problem of the origin of life, the evolution of language and culture, climatic change, the behavior of star clusters, and much more.

Chaos and Complexity

Among the earliest complex systems to be studied were those we now refer to as chaotic. According to chaos theory, the behavior of certain systems can be exceedingly sensitive to the initial conditions, so that systems that start out in very nearly the same state rapidly diverge in their evolution. A classic example is weather forecasting: the slightest error in the input data will grow remorselessly until the forecast and the reality bear little relation to each other. Chaotic systems are therefore unpredictable over the long term.

A chaotic system, even though it may be dynamically very simple and strictly deterministic, may nevertheless behave in a completely random way. For instance, a ball dangled on a string and driven back and forth periodically at the pivot can move unpredictably, circling this way and that without any systematic trend. Such behavior explodes the myth that a world subject to simple laws will necessarily evolve simply. Thus complexity may have its roots in simplicity. On the other hand, chaos (of the form just described) is not anarchy. To coin the phrase, there can be order in chaos. Chaos theory has established that deep mathematical structures may underlie the manner in which a system approaches chaos. It is now recognized that deterministic chaos is not a quirk restricted to a handful of contrived examples but a pervasive feature of dynamical systems, found all around us in nature, from dripping taps through cloud patterns to the orbits of asteroids.

Chaos has certain negative connotations associated with destruction and decay (in Greek mythology *chaos* is the opposite of *cosmos*). However, chaotic systems represent just one case among the much larger class of complex systems. Nonchaotic complexity often has a creative aspect to it. One example concerns the phenomenon of self-organization. Here a system spontaneously leaps from a relatively featureless state to one of greater organized complexity. A famous example is the so-called Bénard instability in fluids. If a pan of water is heated carefully from below, at a critical temperature gradient its initial featureless state will suddenly be transformed into a pattern of hexagonal convection cells—the pattern forms entirely spontaneously. Increasing heating will lead to the breakup of the cells, and chaotic boiling.

Both chaos and self-organization tend to occur in nonlinear systems (roughly, those with strong feedback) that are driven far from equilibrium. Examples have been found in chemical mixtures, biological organisms, collapsing sand piles, ecosystems, star clusters, and much more.

Defining Complexity

The study of complexity is hampered by the lack of a generally accepted definition. To understand the principles that may govern complex systems, there must first be a way of quantifying complexity. Two of this volume's contributors, Gregory Chaitin and Charles Bennett, have attempted precise mathematical definitions of complexity that seek to go beyond the simple designation of "being complex" and characterize subtle but crucial qualities like organization and adaptation. We can all recognize intuitively the distinction between the complexity of a box of gas molecules rushing around at random and the elaborately organized activity of a bacterium. But it's hard to pin down precisely what distinguishes the two. One key fact is clear, though. Organized complexity of the sort exemplified by life doesn't come for free. Nature has to work hard to achieve it; it is forged over deep time, through the elaborate processing of matter and energy and the operation of ratchets to lock in the products of this processing.

Much of the work of Bennett and Chaitin draws on the theory of digital computation for inspiration. For example, Chaitin defines randomness in terms of the input and output states of a computer algorithm. A random distribution of digits is one that is devoid of all patterns, so it cannot be expressed in terms of a more compact algorithm. The definition can be extended to physical objects, too, in terms of the computer algorithm that can simulate or describe them. Chaitin demonstrates that randomness is endemic in mathematics itself and not restricted to physical systems. Quantifying the complexity of a physical system in terms of the difficulty of simulating or describing it with a digital computer opens the way to making precise distinctions between organized and random complexity. It thus provides a natural framework for formulating any emergent laws or principles that might apply to complex systems.

Linking Evolution and Information

At the heart of the computational description of nature is the concept of information. A computer is, after all, an information-processing device that uses a program or algorithm to turn input information into output information. Information arises in physics too, and it does so in several different contexts. In relativity theory, it is information that is forbidden to travel faster than light. In quantum mechanics, the wavefunction represents what we can know about a physical system, that is, its maximal information content. Information also crops up in thermodynamics as the negative of entropy. In biology, information is the key to understanding the nature of life. The living cell resembles a computer in being an information-storing and -processing system: a gene is, after all, a set of instructions for something.

The origin of biological information holds the key to understanding the origin and evolution of life. Can we view evolution as a sort of computer algorithm that generates information step by step, through mutation and selection, or are there other principles at work? Conventional wisdom has it that Darwinian processes are responsible for the information content of genomes; each time natural selection operates to choose one organism over another, information is transferred from the environment into the organism. But two of this volume's contributors challenge this orthodoxy. William Dembski argues that the type of information found in living organisms, namely, complex specified information, cannot arise via an evolutionary algorithm of the Darwinian variety. In particular, he claims that a prebiotic phase of molecular evolution in a chemical soup cannot create anything like a living organism from scratch.

Stuart Kauffman thinks Darwinism may be the truth but not the whole truth. He accepts the Darwinian mechanism for generating some biological complexity but insists it must be augmented by self-organizing principles that can, under some circumstances, be more powerful than Darwinism. This position leads him to address what is perhaps the most baffling property of biological complexity—the well-known ability of living systems to quite literally take on a life of their own and behave as autonomous agents rather than as slaves to the laws of physics and chemistry. How does this come about? How does a physical system harness physics and chemistry to pursue an agenda? Somewhere on the spectrum from a large molecule through bacteria and multicelled organisms to human beings something like purposeful behavior and freedom of choice enters the picture. Complexity reaches a threshold at which the system is liberated from the strictures of physics and chemistry while still remaining subject to their laws. Although the nature of this transition is elusive (Kaufmann makes some specific proposals), its implications for human beings, our perceived place in nature, and issues of free will and ethical responsibility are obvious.

Arrows of Complexity?

Taking a broader perspective, the following question presents itself: is the history of the universe one of slow degeneration toward chaos or a story of the steady emergence of organized complexity? A hundred and fifty years ago Herman von Helmholtz predicted a cosmic heat death based on the second law of thermodynamics, according to which the total entropy of the universe goes on rising until a state of thermodynamic equilibrium is attained, following which little of interest would happen. Thermodynamics thus imprints on the physical world an arrow of time, pointing in the direction of disorder. We certainly see this trend toward chaos and decay all around us in nature, most obviously in the way that the sun and stars are remorselessly consuming their nuclear fuel, presaging their eventual demise in the form of white dwarfs, neutron stars, or black holes.

While it is undeniable that the entropy of the universe is inexorably rising, a summary of cosmological history reveals a very different picture. As I have already remarked, the universe began in a featureless state and has evolved its complexity over time via an extended and elaborate sequence of self-organizing and self-complexifying processes. Among these processes was the emergence of life, its evolution to complex plants and animals, and the subsequent emergence of intelligence, consciousness, and civilization—together with sentient beings who look back on the great cosmic drama and ask what it all means. Thus the cosmic story so far is hardly one of degeneration and decay; rather the most conspicuous indicator of cosmic change is the progressive enrichment of physical systems and the growth of complexity, including organized complexity. So, as Werner Loewenstein and Ian Stewart point out in their respective chapters, there seems to be another arrow of time—a progressive one—pointing in the direction of greater complexity.

Many scientists have suspected there is a general trend, or even a law of nature, that characterizes this advancement of complexity, although a proof is lacking. If such a trend exists, we should like to know whether it involves an additional law or principle of nature—a fourth law of thermodynamics, as Stuart Kauffman and Ian Stewart suggest—or whether it follows from the existing laws of physics.

Stewart considers one of the most-studied examples of the growth of complexity in a cosmological context, namely, gravitation. In many respects a gravitating system seems to go "the wrong way" thermodynamically. For example, a star has a negative specific heat: it gets hotter as it radiates energy, because gravity causes it to shrink. Related to this is the tendency for gravitating systems to grow increasingly clumpy, hence more complex, with time. Contrast the behavior with a gas in a box, in which gravity is neglected. If the initial state is inhomogeneous (e.g., most of the molecules at one end of the box), intermolecular collisions will serve to redistribute the gas toward a uniform state of constant density and pressure. When gravity enters the picture, the evolution of a gas is dramatically different. A large molecular cloud, for example, may start out more or less uniform, but over time it will grow inhomogeneous as star clusters form. On a larger scale, matter is distributed in the form of galaxies, clusters, and superclusters of galaxies. This cosmic structure has evolved over billions of years by gravitational instability and aggregation. Because the growth of large-scale cosmic structure in turn triggered the formation of stars, planets, life, and all the other paraphernalia that adorn the world we see, gravitation may be considered to be the fountainhead of all complexity. Without its curious ability to amplify structure from superficial featurelessness, there would be no sentient beings to write books or discover the laws of gravitation.

A survey of the topics contained in this volume provides the strong impression that the cosmos is poised, exquisitely, between the twin extremes of simplicity and complexity. Too much randomness and chaos would lead to a universe of unstructured anarchy; too much lawlike simplicity would produce regimented uniformity and regularity in which little of interest would occur. The universe is neither a random gas nor a crystal, but a menagerie of coherent, organized, and interacting systems forming a hierarchy of structure. Nature is a thus potent mix of two opposing tendencies, in which there is pervasive spontaneity and novelty, providing openness in the way the universe evolves but enough restraint to impose order on the products. The laws of nature thus bestow on the universe a powerful inherent creativity.

Fine-tuning and Complexity

Is this ability to be creative—to bring forth richness spontaneously—a consequence of very special laws, or is it a generic feature that would arise in almost any universe? My own suspicion is that a creative universe is one in which the laws are delicately fine-tuned to amalgamate simplicity and complexity in a specific manner. This is an example of a set of ideas that has become known, somewhat misleadingly, as "the anthropic principle." It takes as its starting point the accepted fact that the existence of living organisms seems to depend rather sensitively on the fine details of the laws of physics. If those laws were only slightly different, then life, at least of the sort we know, might never have emerged, and the universe would contain no observers to reflect on the significance of the matter.

Suppose, for example, that gravity were slightly stronger, or the nuclear forces slightly weaker—what would be the consequences? How different would the universe be if protons were a tad heavier or if space were four- rather than three-dimensional? What if the big bang had been a little bit bigger? Physicists have studied these topics, and as a general rule they find that the situation in the real universe is rather special. If things were only fractionally different, then stars like the sun would not exist, or heavy elements might never form; in particular, carbon-based life would be unlikely to arise. In short, our existence as observers of the grand cosmic drama seems to hinge on certain delicate mathematical relationships in the underlying laws. It all looks a bit too convenient to be an accident. The British cosmologist Fred Hoyle thought the laws of physics seem so contrived he felt moved to proclaim that the universe is "a put-up job."

What should we make of the apparent coincidence that the laws of physics are "fine-tuned for life"? Does it mean the universe was designed by an intelligent creator with a view to spawning life and mind? Or is the bio-friendly nature of the cosmos just a lucky fluke? Complexity theory can address such matters, because it finds application beyond the description of the physical world, to metaphysics and even theology. Occam's razor encapsulates the general rule that if there are two competing explanations for something, then the one that makes the least number of assumptions is preferred. In other words, all else being equal, simpler explanations are better. One can make this concept more precise by quantifying the complexity of explanations.

Strictly speaking, this is not a scientific question at all but part of metaphysics. The job of the scientist is to take the laws of nature as given—mere brute facts—and get on with the job. But some scientists are not content to accept the existence of an ordered universe as a package of marvels that just happens to be. They ask why the universe is so felicitously bio-friendly and how improbable the existence of such life-encouraging laws might be.

Three responses are traditionally made to this puzzle. The first is to invoke a designer God. Few scientists are prepared to put that sort of theological spin on the facts because it effectively places the explanation beyond the scope of human inquiry. To baldly say "God made it that way" is unlikely to satisfy a skeptical scientist. Others prefer to retreat instead to the cosmic absurdity theory and shrug aside the happy numerical relationships as fortuitous but pointless accidents. They add wryly that if the "coincidences" were not satisfied, we would simply not be here to worry about the significance of it. The trouble with this "so-what?" approach is that it introduces a contradiction into the very foundation of science. The essence of the scientist's belief system is that nature is neither arbitrary nor absurd—there are valid reasons for the way things are. For example, the orbit of the Earth around the sun is shaped like an ellipse because of the inverse square law of gravitation. Since miracles are forbidden in science, there must be a chain of explanation, all the way down to the bedrock of physical reality—the fundamental laws of physics. At this point, cosmic absurdity theorists perform a back-flip and assert that the laws themselves exist reasonlessly: there is no explanation for why the particular laws are as they are. This implies that the universe is ultimately absurd. In other words, we are invited to accept that the logical and rational structure of nature as revealed by science has no absolute anchor. It is grounded in absurdity.

The third, and somewhat more considered, response to the fine-tuning challenge is to appeal to the cosmic lottery theory. The idea here is that there is not just one universe but an enormous collection—sometimes termed "a multiverse"—with each universe coexisting in parallel with the others (or occupying widely separated regions of the same space). The various universes differ in the form of their laws. In some, protons are heavier than neutrons; in others, the strong force is too weak to permit stable heavy nuclei; and so on. Since life depends so delicately on the "right" laws and conditions, it is clear that only a very tiny fraction of these many universes will permit observers. Those beings will then marvel at how contrived and propitious their particular universe seems and how well it suits the emergence of life and mind. But actually they are not the products of beneficent design but merely the winners in a vast and meaningless cosmic lottery.

One objection to the lottery theory is that it still demands laws of some sort and, furthermore, the sort of laws that (at least in one combination) permit something as complex as life and consciousness to happen. A more serious problem, however, is well illuminated by complexity theory. Invoking an infinity of unseen universes just to explain features of the universe we do see seems like overkill—a gross violation of Occam's razor. Lottery theorists retort that unseen universes are better than an unseen God. I suspect, however, that the two explanations are actually equivalent in this regard. What we are trying to explain is how one particular, rather special universe is picked out from an infinity of also-ran contenders. Selecting a candidate from a list is a well-known problem in information theory. Analyzed that way, both theism and the cosmic lottery entail discarding an infinite quantity of information. Both explanations are, in that sense, equally complex. The lottery theory amounts to little more than theism dressed in science's clothing. I suspect that a proper application of algorithmic complexity theory to these apparently competing explanations would demonstrate their formal equivalence.

But some further subtle issues are involved. The purpose of an explanation for something is normally to account for it in terms of something simpler, or more fundamental. If one is trying to explain the physical universe in terms of God, the question arises of whether the creator is more or less complex than the creature. If God is more complex than the universe, skeptics will argue that nothing is gained by invoking God. Richard Dawkins has made this point forcefully. Richard Swinburne, however, has countered that God is indeed simple and therefore is a preferred economic explanation to accepting a complex universe as a brute fact.

Complexity Studies and the Quest for Meaning

Following the lead of complexity studies, a substantial part of current science can be described as the study of compressions. The claim is that compressed algorithms may give us a clue for understanding how complex systems are generated and propagated by relatively simple algorithms that are able to incorporate the overwhelming fluffiness of environmental conditions.

However, complexity theory is more than a research paradigm: It is also an incentive for an emergentist worldview that impinges on the question of the meaning of the universe. Even though most would agree that meaning is not part of the agenda of science itself, "why" questions may well motivate research; furthermore, scientific discoveries will inevitably give new twists to the perennial quest for some sort of ultimate meaning or purpose. It is a baffling fact that even if the causal *route* from complexity to life can be explained in the prosaic language of mathematics, the *outcome* of these prosaic processes indeed does evoke poetic descriptions. Out of the simple arises complexity, a vibrant world of life and sentience, joy and suffering. Science as such does not care much about the question of the meaning or meaninglessness of natural systems. But our responses to the quest for cosmic meaning are inevitably shaped by scientific assumptions concerning the way the world is and how nature works.

In this sense the scientific quest for explaining the route "from complexity to life" provokes a postscientific quest for understanding the emergence of meaning, "from complexity to consciousness." This is where philosophy and theology may enter the picture. If by philosophical theology we mean the disciplined attempt to rationally evaluate alternative philosophical and religious answers to the question of meaning in a world of emerging complexity, what then are the options available? One way to go would be to stay with the idea of God as the ingenious Architect of the world, who has contrived the laws of nature so that sentient and intelligent beings (such as us) will arise to reflect on the wonder of it all. This notion of God as a metadesigner provides an obvious, if somewhat simplistic, theological basis to the scientific notion of human beings "at home in the universe."

As long as the designer God is seen as selecting the basic laws of physics, this concept of design is fully compatible with a naturalistic understanding of the world. In fact, the so-called anthropic principle was developed neither by philosophers nor by theologians but by cosmologists such as Brandon Carter, Martin Rees, and Bernard Carr in the 1970s. As mentioned earlier, however, design theism is not required by the existence of cosmic coincidences; other explanations exist. In addition, the design argument is fairly generic in nature since it relates only to the basic laws of physics that constitute the general setup of the universe. Apart from the cosmic selector function, then, God's nature remains mysterious in this picture.

Some theists want to take the idea of design a step further, using "intelligent design" to explain not just the underlying laws of physics but certain states of the world too, specifically, living organisms. The central claim, propounded in this volume by William Dembski, is that life at the cellular level manifests a form of irreducible complexity that defies any naturalistic explanation.

As argued by Niels Henrik Gregersen in this volume, this version of the design argument differs markedly from the design hypothesis in relation to the anthropic principle. Whereas "intelligent design" is explicitly antinaturalistic and seeks to replace the paradigm of self-organization by reference to an intelligent designer, the theistic interpretation of the anthropic coincidences presupposes an emergentist monism, based on no entities other than the material elements known or knowable through physics. As expressed by Arthur Peacocke, the only dualism permitted by a theistic naturalism is the distinction between the order of creation and God the creator, understood as the ultimate basis of existence. On this view, there is no reason to see a conflict between God's creativity and nature's capacity for self-organization.

However one wants to position oneself in this debate, it should be noted that major strands of modern theology remain altogether skeptical about the Enlightenment notion of a purely transcendent designer

God. In particular, the wide majority of modern theologians find the intelligent design proposal unattractive, both because it seems to endanger the God-given autonomy of the natural order and because it seems to commit the fallacy of misplaced concreteness. God becomes too closely tied up with (assumed) gaps in scientific explanation. Accordingly, God's activity seems to be confined to the role of a cosmic magician who overrules the created order, attributable to God in the first place, rather than that of the beneficent creator who supports the in-built capacities of matter.

According to Harold Morowitz, an emergentist worldview suggests that the divine immanence unfolds into the domain of an evolving world that makes transcendence possible. On this view, the transcendence of God is itself an emergent reality. God is not only the wellspring of the natural world but also a result that flows out of the workings of nature. The human mind is then God's transcendence, God's image. This idea of God as an emergent reality of evolution would probably imply a radical reinterpretation of religious tradition.

Another option is suggested by Arthur Peacocke, who argues for a so-called panentheistic model. God is here seen both as ontologically distinct from the world (and thus as necessarily transcendent) and yet as immanent in, with, and under natural processes. On this model, creation is seen as the self-expression of God the Word or Logos. God is equally expressed everywhere in the universe, not so much in the dull inertness of matter as in the complex beauty of the universe, not so much in selfishness as in self-giving love. Important for this position is the question whether emergent realities can themselves be causally efficacious and further propagate to enrich the cosmic order. For to be "real, new and irreducible . . . must be to have new irreducible causal powers," according to Samuel Alexander, an early proponent of emergenticism.

The emergentist worldview seems to present us with a twofold task requiring a collaboration between the natural sciences, philosophy, and theology. The first is about the causal structure of our world. It seems that the universe is driven by different sorts of causality. If, as suggested by Stuart Kauffman in this volume, lawlike tendencies toward complexity inevitably end up producing autonomous agents that are able to perform various thermodynamical work cycles, and if the concrete movements and whereabouts of these autonomous agents are not predetermined by the general laws that produced them in the first place, then we shall never be in a position to prestate all possible adaptions in the history of evolution. The causal structure of the universe would then consist of an intricate interplay between the "constitutive" and global laws of physics and the local "structuring" constraints that are exercised via the specific informational states of evolved agents. The atmosphere surrounding our planet is itself a result of myriads of such thermodynamical work cycles; indeed, local autonomous agents performing photosynthesis have influenced the biosphere as a whole. Analogous sorts of multilevel causality may also be at work in human beings. The functioning of our brains is a constitutive cause of our sentience and thoughts; yet how we use our brains is codetermined by the language and behavior of sociocultural agents; thus our concrete mental operations (such as doing math or playing the violin eight hours daily) will have a feedback influence on the neural structure of the individual brains. There seem to be different sorts of causality at work here. Some speak of bottom-up versus top-down causality, while other speaks of constitutive versus structuring causality. Regardless of our own religious sympathies or antipathies, we approach much of life through emergent qualities such as trust, love, and the sense of beauty. If these phenomena now have a safer place in the causal fabric of reality, they can no longer be deemed to be mere epiphenomena. This new emphasis on top-down or informational causality will probably also influence the manner in which theologians conceive of God's interaction with the evolving world of autonomous agents.

The second question relates to meaning: How does a sense of meaning emerge from a universe of inanimate matter subject to blind and purposeless forces? Perhaps the lesson to be learned from complexity studies (a feature that links it so well with evolutionary theory) is that nature is not only a self-repetitive structure but a structure that seems as if it is geared for letting specified autonomous agents appear and propagate further. If this is so, we need both scientific explanations of the general principles underlying natural processes and accounts that are sensitive to the specifics and capable of explaining to us why we have evolved to be the particular creatures we are today.

We seem to be constructed from an overwhelmingly vast tapestry of biological possibilities, yet our lives also consist of singularities. The hard mysteries of existence are no longer placed only in the very small (in quantum physics) and in the vastness of the universe but also in the realm of the exceedingly complex. A high degree of complexity always implies a high degree of specificity: a "thisness," or haecceity, as the medieval philosopher Duns Scotus called it. At a personal level, we may tend to regard our lives as a gift or as a burden. But in the final analysis the question of meaning is about how to come to terms with the specificity of our individual existence.

PART I

Defining Complexity

TWO

RANDOMNESS AND
MATHEMATICAL PROOF

Gregory J. Chaitin

Although randomness can be precisely defined and can even be measured, a given number cannot be proved to be random. This enigma establishes a limit to what is possible in mathematics.

Almost everyone has an intuitive notion of what a random number is. For example, consider these two series of binary digits:

> 01010101010101010101

> 01101100110111100010

The first is obviously constructed according to a simple rule; it consists of the number 01 repeated 10 times. If one were asked to speculate on how the series might continue, one could predict with considerable confidence that the next two digits would be 0 and 1. Inspection of the second series of digits yields no such comprehensive pattern. There is no obvious rule governing the formation of the number, and there is no rational way to guess the succeeding digits. The arrangement seems haphazard; in other words, the sequence appears to be a random assortment of 0s and 1s.

The second series of binary digits was generated by flipping a coin 20 times and writing a 1 if the outcome was heads and a 0 if it was tails. Tossing a coin is a classical procedure for producing a random number, and one might think at first that the provenance of the series alone would certify that it is random. This is not so. Tossing a coin 20 times can produce any one of 2^{20} (or a little more than a million) binary series, and each of them has exactly the same probability. Thus it should be no more surprising to obtain the series with an obvious pattern than to obtain the one that seems to be random; each represents an event with a probability of 2^{-20}. If origin in a probabilistic event were made the sole criterion of randomness, then both series would have to be considered random, and indeed so would all others, since the same mechanism can generate all the possible series. The conclusion is singularly unhelpful in distinguishing the random from the orderly.

Clearly a more sensible definition of randomness is required, one that does not contradict the intuitive concept of a "patternless" number. Such a definition has been devised only in the past few decades. It does not consider the origin of a number but depends entirely on the characteristics of the sequence of digits. The new definition enables us to describe the properties of a random number more precisely than was formerly possible, and it establishes a hierarchy of degrees of randomness. Of perhaps even greater interest than the capabilities of the definition, however, are its limitations. In particular, the definition cannot help to determine, except in very special cases, whether or not a given series of digits, such as the second one just given, is in fact random or only seems to be random (Chaitin 1974). This limitation is not a flaw in the definition; it is a consequence of a subtle but fundamental anomaly in the foundation of mathematics. It is closely related to a famous theorem devised and proved in 1931 by Kurt Gödel, which has come to be known as Gödel's incompleteness theorem. Both the theorem and the recent discoveries concerning the nature of randomness help to define the boundaries that constrain certain mathematical methods.

Algorithmic Definition

The new definition of randomness has its heritage in information theory, the science, developed mainly since World War II, that studies the transmission of messages. Suppose you have a friend who is visiting a planet in another galaxy, and sending him telegrams is very expensive. He forgot to take along his tables of trigonometric functions and he has asked you to supply them. You could simply translate the numbers into an appropriate code (such as the binary numbers) and transmit them directly, but even the most modest tables of the six functions have a few thousand digits, so the cost would be high. A much cheaper way to convey the same information would be to transmit instructions for calculating the tables from the underlying trigonometric formulas, such as Euler's equation $e^{ix} = \cos x + i \sin x$. Such a message could be relatively brief, yet inherent in it is all the information contained in even the largest tables.

Suppose, on the other hand, your friend is interested not in trigonometry but in baseball. He would like to know the scores of all the major-league games played since he left the Earth some thousands of years before. In this case it is most unlikely that a formula could be found for compressing the information into a short message; in such a series of numbers each digit is essentially an independent item of information, and it cannot be predicted from its neighbors or from some underlying rule. There is no alternative to transmitting the entire list of scores.

In this pair of whimsical messages is the germ of a new definition of randomness. It is based on the observation that the information embodied in a random series of numbers cannot be "compressed," or reduced to a more compact form. In formulating the actual definition it is preferable to consider communication not with a distant friend but with a digital computer. The friend might have the wit to make inferences about numbers or to construct a series from partial information or from vague instructions. The computer does not have that capacity, and for our purposes that deficiency is an advantage. Instructions given the computer must be complete and explicit, and they must enable it to proceed step by step without requiring that it comprehend the result of any part of the operations it performs. Such a program of instructions is an algorithm. It can demand any finite number of mechanical manipulations of numbers, but it cannot ask for judgments about their meaning.

The definition also requires that we be able to measure the information content of a message in some more precise way than by the cost of sending it as a telegram. The fundamental unit of information is the "bit," defined as the smallest item of information capable of indicating a choice between two equally likely things. In binary notation one bit is equivalent to one digit, either a 0 or a 1.

We are now able to describe more precisely the differences between the two series of digits presented at the beginning of this chapter:

$$01010101010101010101$$

$$01101100110111100010$$

The first could be specified to a computer by a very simple algorithm, such as "Print 01 ten times." If the series were extended according to the same rule, the algorithm would have to be only slightly larger; it might be made to read, for example, "Print 01 a million times." The number of bits in such an algorithm is a small fraction of the number of bits in the series it specifies, and as the series grows larger the size of the program increases at a much slower rate.

For the second series of digits there is no corresponding shortcut. The most economical way to express the series is to write it out in full, and the shortest algorithm for introducing the series into a computer would be "Print 01101100110111100010." If the series were much larger (but still apparently patternless), the algorithm would have to be expanded to the corresponding size. This "incompressibility" is a property of all random numbers; indeed, we can proceed directly to define randomness in terms of incompressibility: A series of numbers is random if the smallest algorithm capable of specifying it to a computer has about the same number of bits of information as the series itself.

This definition was independently proposed about 1965 by A. N. Kolmogorov of the Academy of Science of the USSR and by me (Chaitin 1966, 1969), when I was an undergraduate at the City College of the City University of New York. Both Kolmogorov and I were then unaware of related proposals made in 1960 by Ray J. Solomonoff of the Zator Company in an endeavor to measure the simplicity of scientific theories. In the past few decades we and others have continued to explore the meaning of randomness. The original formulations have been improved, and the feasibility of the approach has been amply confirmed.

Model of Inductive Method

The algorithmic definition of randomness provides a new foundation for the theory of probability. By no means does it supersede classical probability theory, which is based on an ensemble of possibilities, each of which is assigned a probability. Rather, the algorithmic approach complements the ensemble method by giving precise meaning to concepts that had been intuitively appealing but could not be formally adopted.

The ensemble theory of probability, which originated in the seventeenth century, remains today of great practical importance. It is the foundation of statistics, and it is applied to a wide range of problems in science and engineering. The algorithmic theory also has important implications, but they are primarily theoretical. The area of broadest interest is its amplification of Gödel's incompleteness theorem. Another application (which actually preceded the formulation of the theory itself) is in Solomonoff's model of scientific induction.

Solomonoff represented a scientist's observations as a series of binary digits. The scientist seeks to explain these observations through a theory, which can be regarded as an algorithm capable of generating the series and extending it, that is, predicting future observations. For any given series of observations there are always several competing theories, and the scientist must choose among them. The model demands that the smallest algorithm, the one consisting of the fewest bits, be selected. Put another way, this rule is the familiar formulation of Occam's razor: Given differing theories of apparently equal merit, the simplest is to be preferred.

Thus in the Solomonoff model a theory that enables one to understand a series of observations is seen as a small computer program that reproduces the observations and makes predictions about possible future observations. The smaller the program, the more comprehensive the theory and the greater the degree of understanding. Observations that are random cannot be reproduced by a small program and therefore cannot be explained by a theory. In addition, the future behavior of a random system cannot be predicted. For random data the most compact way for the scientist to communicate the observations is to publish them in their entirety.

Defining randomness or the simplicity of theories through the capabilities of the digital computer would seem to introduce a spurious element into these essentially abstract notions: the peculiarities of the particular computing machine employed. Different machines communicate through different computer languages, and a set of instructions expressed in one of those languages might require more or fewer bits when the instructions are translated into another language. Actually, however, the choice of computer matters very little. The problem can be avoided entirely simply by insisting that the randomness of all numbers be tested on the same machine. Even when different machines are employed, the idiosyncrasies of various languages can readily be compensated for. Suppose, for example, someone has a program written in English and wishes to utilize it with a computer that reads only French. Instead of translating the algorithm itself he could preface the program with a complete English course written in French. Another mathematician with a French program and an English machine would follow the opposite procedure. In this way only a fixed number of bits need be added to the program, and that number grows less significant as the size of the series specified by the program increases. In practice a device called a compiler often makes it possible to ignore the differences between languages when one is addressing a computer.

Since the choice of a particular machine is largely irrelevant, we can choose for our calculations an ideal computer. It is assumed to have unlimited storage capacity and unlimited time to complete its calculations. Input to and output from the machine are both in the form of binary digits. The machine begins to operate as soon as the program is given it, and it continues until it has finished printing the binary series that is the result. The machine then halts. Unless an error is made in the program, the computer will produce exactly one output for any given program.

Minimal Programs and Complexity

Any specified series of numbers can be generated by an infinite number of algorithms. Consider, for example, the three-digit decimal series 123. It could be produced by an algorithm such as "Subtract 1 from 124 and print the result," or "Subtract 2 from 125 and print the result," or an infinity of other programs formed on the same model. The programs of greatest interest, however, are the smallest ones that will yield a given numerical series. The smallest programs are called minimal programs; for a given series there may be only one minimal program or there may be many.

Any minimal program is necessarily random, whether or not the series it generates is random. This conclusion is a direct result of the way we have defined randomness. Consider the program P, which is a minimal program for the series of digits S. If we assume that P is not random, then by definition there must be another program, P', substantially smaller than P, that will generate it. We can then produce S by the following algorithm: "From P' calculate P, then from P calculate S." This program is only a few bits longer than P' and thus it must be substantially shorter than P. P is therefore not a minimal program.

The minimal program is closely related to another fundamental concept in the algorithmic theory of randomness: the concept of complexity. The complexity of a series of digits is the number of bits that must be put into a computing machine in order to obtain the original series as output. The complexity is therefore equal to the size in bits of the minimal programs of the series. Having introduced this concept, I can now restate the definition of randomness in more rigorous terms:

A random series of digits is one whose complexity is approximately equal to its size in bits.

The notion of complexity serves not only to define randomness but also to measure it. Given several series of numbers each having n digits, it is theoretically possible to identify all those of complexity $n-1$, $n-10$, $n-100$, and so forth and thereby to rank the series in decreasing order of randomness. The exact value of complexity below which a series is no longer considered random remains somewhat arbitrary. The value ought to be set low enough for numbers with obviously random properties not to be excluded and high enough for numbers with a conspicuous pattern to be disqualified, but to set a particular numerical value is to judge what degree of randomness constitutes actual randomness. It is this uncertainty that is reflected in the qualified statement that the complexity of a random series is approximately equal to the size of the series.

Properties of Random Numbers

The methods of the algorithmic theory of probability can illuminate many of the properties of both random and nonrandom numbers. The frequency distribution of digits in a series, for example, can be shown to have an important influence on the randomness of the series. Simple inspection suggests that a series consisting entirely of either 0s or 1s is far from random, and the algorithmic approach confirms that conclusion. If such a series is n digits long, its complexity is approximately equal to the logarithm to the base 2 of n. (The exact value depends on the machine language employed.) The series can be produced by a simple algorithm such as "Print 0 n times," in which virtually all the information needed is contained in the binary numeral for n. The size of this number is about $\log_2 n$ bits. Since for even a moderately long series the logarithm of n is much smaller than n itself, such numbers are of low complexity; their intuitively perceived pattern is mathematically confirmed.

Another binary series that can be profitably analyzed in this way is one where 0s and 1s are present with relative frequencies of three-fourths and one-fourth. If the series is of size n, it can be demonstrated that its complexity is no greater than four-fifths n, that is, a program that will produce the series can be written in $4n/5$ bits. This maximum applies regardless of the sequence of the digits, so that no series with such a frequency distribution can be considered very random. In fact, it can be proved that in any long binary series that is random the relative frequencies of 0s and 1s must be very close to one-half. (In a random decimal series the relative frequency of each digit is, of course, one-tenth.)

Numbers having a nonrandom frequency distribution are exceptional. Of all the possible n-digit binary numbers there is only one, for example, that consists entirely of 0s and only one that is all 1s. All the rest are less orderly, and the great majority must, by any reasonable standard, be called random. To choose an arbitrary limit, we can calculate the fraction of all n-digit binary numbers that have a complexity of less than $n-10$. There are 2^1 programs one digit long that might generate an n-digit series; there are 2^2 programs two digits long that could yield such a series, 2^3 programs three digits long, and so forth, up to the longest programs permitted within the allowed complexity; of these there are 2^{n-11}. The sum of this series $(2^1 + 2^2 + \ldots + 2^{n-11})$ is equal to $2^{n-10} - 2$. Hence there are fewer than 2^{n-10} programs of size less than $n-10$, and since each of these programs can specify no more than one series of digits, fewer than 2^{n-10} of the 2^n numbers have a complexity less than $n-10$. Since $2^{n-10} / 2^n = 1/1,024$, it follows that of all the n-digit binary numbers only about one in 1,000 have a complexity less than n-10. In other words, only about one series in 1,000 can be compressed into a computer program more than 10 digits smaller than itself.

A necessary corollary of this calculation is that more than 999 of every 1,000 n-digit binary numbers have a complexity equal to or greater than $n-10$. If that degree of complexity can be taken as an appropriate test of randomness, then almost all n-digit numbers are in fact random. If a fair coin is tossed n times, the probability is greater than .999 that the result will be random to this extent. It would therefore seem easy to exhibit a specimen of a long series of random digits; actually it is impossible to do so.

Formal Systems

It can readily be shown that a specific series of digits is not random; it is sufficient to find a program that will generate the series and that is substantially smaller than the series itself. The program need not be a minimal program for the series; it need only be a small one. To demonstrate that a particular series of digits is random, on the other hand, one must prove that no small program for calculating it exists.

It is in the realm of mathematical proof that Gödel's incompleteness theorem is such a conspicuous landmark; my version of the theorem predicts that the required proof of randomness cannot be found. The consequences of this fact are just as interesting for what they reveal about Gödel's theorem as for what they indicate about the nature of random numbers.

Gödel's theorem represents the resolution of a controversy that preoccupied mathematicians during the early years of the twentieth century. The question at issue was: "What constitutes a valid proof in mathematics and how is such a proof to be recognized?" David Hilbert had attempted to resolve the controversy by devising an artificial language in which valid proofs could be found mechanically, without any need for human insight or judgment. Gödel showed that there is no such perfect language.

Hilbert established a finite alphabet of symbols; an unambiguous grammar specifying how a meaningful statement could be formed; a finite list of axioms, or initial assumptions; and a finite list of rules of inference for deducing theorems from the axioms or from other theorems. Such a language, with its rules, is called a formal system.

A formal system is defined so precisely that a proof can be evaluated by a recursive procedure involving only simple logical and arithmetical manipulations. In other words, in the formal system there is an algorithm for testing the validity of proofs. Today, although not in Hilbert's time, the algorithm could be executed on a digital computer and the machine could be asked to "judge" the merits of the proof.

Because of Hilbert's requirement that a formal system have a proof-checking algorithm, it is possible in theory to list one by one all the theorems that can be proved in a particular system. One first lists in alphabetical order all sequences of symbols one character long and applies the proof-testing algorithm to each of them, thereby finding all theorems (if any) whose proofs consist of a single character. One then tests all the two-character sequences of symbols, and so on. In this way all potential proofs can be checked, and eventually all theorems can be discovered in order of the size of their proofs. (The method is, of course, only a theoretical one; the procedure is too lengthy to be practical.)

Unprovable Statements

Gödel showed in his 1931 proof that Hilbert's plan for a completely systematic mathematics cannot be fulfilled. He did this by constructing an assertion about the positive integers in the language of the formal system that is true but that cannot be proved in the system. The formal system, no matter how large or how carefully constructed it is, cannot encompass all true theorems and is therefore incomplete. Gödel's technique can be applied to virtually any formal system, and it therefore demands the surprising and, for many, discomforting conclusion that there can be no definitive answer to the question "What is a valid proof?" Gödel's proof of the incompleteness theorem is based on the paradox of Epimenides the Cretan, who is said to have averred, "All Cretans are liars" (see Quine 1962). The paradox can be rephrased in more general terms as "This statement is false," an assertion that is true if and only if it is false and that is therefore neither true nor false. Gödel replaced the concept of truth with that of provability and thereby constructed the sentence "This statement is unprovable" an assertion that, in a specific formal system, is provable if and only if it is false. Thus either a falsehood is provable, which is forbidden, or a true statement is unprovable, and hence the formal system is incomplete. Gödel then applied a technique that uniquely numbers all statements and proofs in the formal system and thereby converted the sentence "This statement is unprovable" into an assertion about the properties of the positive integers. Because this transformation is possible, the incompleteness theorem applies with equal cogency to all formal systems in which it is possible to deal with the positive integers (see Nagel & Newman 1956).

The intimate association between Gödel's proof and the theory of random numbers can be made plain through another paradox, similar in form to the paradox of Epimenides. It is a variant of the Berry paradox, first published in 1908 by Bertrand Russell. It reads: "Find the smallest positive integer which to be specified requires more characters than there are in this sentence." The sentence has 114 characters (counting spaces between words and the period but not the quotation marks), yet it supposedly specifies an integer that, by definition, requires more than 114 characters to be specified.

As before, in order to apply the paradox to the incompleteness theorem it is necessary to remove it from the realm of truth to the realm of provability. The phrase "which requires" must be replaced by "which can be proved to require," it being understood that all statements will be expressed in a particular formal system. In addition the vague notion of "the number of characters required to specify" an integer can be replaced by the precisely defined concept of complexity, which is measured in bits rather than characters.

The result of these transformations is the following computer program: "Find a series of binary digits that can be proved to be of a complexity greater than the number of bits in this program." The program tests all possible proofs in the formal system in order of their size until it encounters the first one proving that a specific binary sequence is of a complexity greater than the number of bits in the program. Then it prints the series it has found and halts. Of course, the paradox in the statement from which the program was derived has not been eliminated. The program supposedly calculates a number that no program its size should be able to calculate. In fact, the program finds the first number that it can be proved incapable of finding.

The absurdity of this conclusion merely demonstrates that the program will never find the number it is designed to look for. In a formal system one cannot prove that a particular series of digits is of a complexity greater than the number of bits in the program employed to specify the series.

A further generalization can be made about this paradox. It is not the number of bits in the program itself that is the limiting factor but the number of bits in the formal system as a whole. Hidden in the program are the axioms and rules of inference that determine the behavior of the system and provide the algorithm for testing proofs. The information content of these axioms and rules can be measured and can be designated the complexity of the formal system. The size of the entire program therefore exceeds the complexity of the formal system by a fixed number of bits c. (The actual value of c depends on the machine language employed.) The theorem proved by the paradox can therefore be put as follows: In a formal system of complexity n it is impossible to prove that a particular series of binary digits is of complexity greater than $n + c$, where c is a constant that is independent of the particular system employed.

Limits of Formal Systems

Since complexity has been defined as a measure of randomness, this theorem implies that in a formal system no number can be proved to be random unless the complexity of the number is less than that of the system itself. Because all minimal programs are random the theorem also implies that a system of greater complexity is required in order to prove that a program is a minimal one for a particular series of digits.

The complexity of the formal system has such an important bearing on the proof of randomness because it is a measure of the amount of information the system contains and hence of the amount of information that can be derived from it. The formal system rests on axioms: fundamental statements that are irreducible in the same sense that a minimal program is. (If an axiom could be expressed more compactly, then the briefer statement would become a new axiom and the old one would become a derived theorem.) The information embodied in the axioms is thus itself random, and it can be employed to test the randomness of other data. The randomness of some numbers can therefore be proved but only if they are smaller than the formal system. Moreover, any formal system is of necessity finite, whereas any series of digits can be made arbitrarily large. Hence there will always be numbers whose randomness cannot be proved.

The endeavor to define and measure randomness has greatly clarified the significance and the implications of Gödel's incompleteness theorem. That theorem can now be seen not as an isolated paradox but as a natural consequence of the constraints imposed by information theory. Hermann Weyl said that the doubt induced by such discoveries as Gödel's theorem had been "a constant drain on the enthusiasm and determination with which I pursued my research work" (1946, p. 13). From the point of view of information theory, however, Gödel's theorem does not appear to give cause for depression. Instead it seems simply to suggest that in order to progress, mathematicians, like investigators in other sciences, must search for new axioms.

Illustrations

(a) $10100 \rightarrow$ Computer $\rightarrow$ 11111111111111111111
(b) $01101100110111100010 \rightarrow$ Computer $\rightarrow$ 01101100110111100010
Algorithmic definition of randomness relies on the capabilities and limitations of the digital computer. In order to produce a particular output, such as a series of binary digits, the computer must be given a set of explicit instructions that can be followed without making intellectual judgments. Such a program of instructions is an algorithm. If the desired output is highly ordered (a), a relatively small algorithm will suffice; a series of 20 1s, for example, might be generated by some hypothetical computer from the program 10100, which is the binary notation for the decimal number 20. For a random series of digits (b) the most concise program possible consists of the series itself. The smallest programs capable of generating a particular series are called the minimal programs of the series; the size of these programs, mea-

sured in bits, or binary digits, is the complexity of the series. A series of digits is defined as random if series' complexity approaches its size in bits.

Alphabet, Grammar, Axioms, Rules of Inference

↓

Computer

↓

Theorem 1, Theorem 2, Theorem 3, Theorem 4, Theorem 5, . . .

Formal systems devised by David Hilbert contain an algorithm that mechanically checks the validity of all proofs that can be formulated in the system. The formal system consists of an alphabet of symbols in which all statements can be written; a grammar that specifies how the symbols are to be combined; a set of axioms, or principles accepted without proof; and rules of inference for deriving theorems from the axioms. Theorems are found by writing all the possible grammatical statements in the system and testing them to determine which ones are in accord with the rules of inference and are therefore valid proofs. Since this operation can be performed by an algorithm it could be done by a digital computer. In 1931 Kurt Gödel demonstrated that virtually all formal systems are incomplete: in each of them there is at least one statement that is true but that cannot be proved.

- Observations: 0101010101
- Predictions: 01010101010101010101
- Theory: Ten repetitions of 01
- Size of Theory: 21 characters
- Predictions: 01010101010000000000
- Theory: Five repetitions of 01 followed by ten 0s
- Size of Theory: 41 characters

Inductive reasoning as it is employed in science was analyzed mathematically by Ray J. Solomonoff. He represented a scientist's observations as a series of binary digits; the observations are to be explained and new ones are to be predicted by theories, which are regarded as algorithms instructing a computer to reproduce the observations. (The programs would not be English sentences but binary series, and their size would be measured not in characters but in bits.) Here two competing theories explain the existing data; Occam's razor demands that the simpler, or smaller, theory be preferred. The task of the scientist is to search for minimal programs.

If the data are random, the minimal programs are no more concise than the observations and no theory can be formulated.

Illustration is a graph of number of n-digit sequences as a function of their complexity. The curve grows exponentially from approximately 0 to approximately 2^n as the complexity goes from 0 to n.

Random sequences of binary digits make up the majority of all such sequences. Of the 2^n series of n digits, most are of a complexity that is within a few bits of n. As complexity decreases, the number of series diminishes in a roughly exponential manner. Orderly series are rare; there is only one, for example, that consists of n 1s.

- Russell Paradox: Consider the set of all sets that are not members of themselves. Is this set a member of itself?
- Epimenides Paradox: Consider this statement: "This statement is false." Is this statement true?
- Berry Paradox: Consider this sentence: "Find the smallest positive integer which to be specified requires more characters than there are in this sentence." Does this sentence specify a positive integer?

Three paradoxes delimit what can be proved. The first, devised by Bertrand Russell, indicated that informal reasoning in mathematics can yield contradictions, and it led to the creation of formal systems. The second, attributed to Epimenides, was adapted by Gödel to show that even within a formal system there are true statements that are unprovable. The third leads to the demonstration that a specific number cannot be proved random.

(a) This statement is unprovable.
(b) The complexity of 01101100110111100010 is greater than 15 bits.
(c) The series of digits 01101100110111100010 is random.
(d) 10100 is a minimal program for the series 11111111111111111111.

Unprovable statements can be shown to be false, if they are false, but they cannot be shown to be true. A proof that "This statement is unprovable" (a) reveals a self-contradiction in a formal system. The assignment of a numerical value to the complexity of a particular number (b) requires a proof that no smaller algorithm for generating the number exists; the proof could be supplied only if the formal system itself were more complex than the number. Statements labeled (c) and (d) are subject to the same limitation, since the identification of a random number or a minimal program requires the determination of complexity.

REFERENCES

Chaitin,Gregory J. 1966. "On the Length of Programs for Computing Finite Binary Sequences."*Journal of the Association of Computing Machinery* 13: 547–569.

Chaitin,Gregory J. 1969. "On the Length of Programs for Computing Finite Binary Sequences: Statistical Considerations." *Journal of the Association for Computing Machinery* 16: 145–159

Chaitin, Gregory J. 1974. "Information-Theoretic Limitations of Formal Systems." *Journal of the Association for Computing Machinery* 21: 403–424.

Gödel, Kurt. 1931. "Über formal unentscheidbare Sätze der Principia Mathematica und verwandter Systeme I." *Monatshefte für Mathematik und Physik* 38: 173–198.

Kolmogorov, Andre N. 1965. "Three Approaches to the Quantitive Definition of Information." *Problems of Information Transmission* 1: 1–17.

Nagel, Ernest and James R. Newman. 1956. "Gödel's Proof." *Scientific American* 194(6): 71–86.

Quine, W. V. 1962. "Paradox." *Scientific American* 206(4): 84–96.

Russell, Bertrand. 1908. "Mathematical Logic as Based on the Theory of Types." *American Journal of Mathematics* 30: 222–262. Reprinted in Jean van Heijenoort, *From Frege to Gödel.* Cambridge, Mass.: Harvard University Press, 1967: 152–182.

Solomonoff, R. J. 1960. "A Preliminary Report on a General Theory of Inductive Inference." Report ZTB-138, Zator Company, Cambridge, Mass.

Weyl, Hermann. 1946. "Mathematics and Logic." *American Mathematical Monthly* 53: 2–13.

FURTHER READING

Casti, John and Werner Depauli. 2000. *Gödel: A Life of Logic.* Cambridge, Mass.: Perseus.

Chaitin, Gregory J. 1998, 1999. *The Limits of Mathematics, The Unknowable.* Singapore: Springer.

Chaitin, Gregory J. 2001, 2002. *Exploring Randomness, Conversations with a Mathematician.* London: Springer.

Grattan-Guinness, Ivor. 2000. *The Search for Mathematical Roots, 1870–1940.* Princeton: Princeton University Press.

Ruelle, David. 1991. *Chance and Chaos,* Princeton: Princeton University Press.

Tasić, Vladimir. 2001. *Mathematics and the Roots of Postmodern Thought.* New York: Oxford University Press.

Tymoczko, Thomas. 1998. *New Directions in the Philosophy of Mathematics.* Princeton: Princeton University Press.

THREE

HOW TO DEFINE COMPLEXITY
IN PHYSICS, AND WHY

Charles H. Bennett

Various notions of complexity are listed and discussed in this chapter. The advantage of having a definition of complexity that is rigorous and yet in accord with intuitive notions is that it allows certain complexity-related questions in statistical physics and the theory of computation to be posed well enough to be amenable to proof or refutation.

Natural irreversible processes are nowadays thought to have a propensity for self-organization—the spontaneous generation of complexity. One may attempt to understand the origin of complexity in serveral ways. One can attempt to elucidate the actual course of galactic, solar, terrestrial, biological, and even cultural evolution. One can attempt to make progress on epistemological questions such as the anthropic principle (Bennett 1982)—the ways in which the complexity of the universe is conditioned by the existence of sentient observers—and the question often raised in connection with interpretations of quantum mechanics of what, if any, distinction science should make between the world that did happen and the possible worlds that might have happened. One can seek a cosmological "theory of everything" without which it would seem no truly general theory of natural history can be built. Finally, at an intermediate level of humility, one can attempt to discover general principles governing the creation and destruction of complexity in the standard mathematical models of many-body systems, for example, stochastic cellular automata such as the Ising model, and partial differential equations such as those of hydrodynamics or chemical reactions-diffusion. An important part of the latter endeavor is the formulation of suitable definitions of complexity: definitions that on the one hand adequately capture intuitive notions of complexity and on the other hand are sufficiently objective and mathematical to prove theorems about. In what follows I list and comment on several candidates for a complexity measure in physics, advocating on "logical depth" as most suitable for the development of a general theory of complexity in many-body systems. Further details can be found in Bennett (1987).

How: Candidates for a Satisfactory Formal Measure of Complexity

Life-like Properties

Life-like properties (e.g., growth, reproduction, adaption) are very hard to define rigorously, and are too dependent on function, as opposed to structure. Intuitively, a dead human body is still complex, though it is functionally inert.

Thermodynamic Potentials

Thermodynamic potentials (entropy, free energy) measure a system's capacity for irreversible change but do not agree with intuitive notions of complexity. For example, a bottle of sterile nutrient solution has higher free energy, but lower subjective complexity, than the bacterial culture it would turn into if inocculated with a single bacterium. The rapid growth of bacteria following introduction of a seed bacterium is a thermodynamically irreversible process analogous to crystalization of a supersaturated solution following introduction of a seed crystal. Even without the seed either of these processes is vastly more probable than its reverse: spontaneous melting of crystal into supersaturated solution or transformation of bacteria into high-free-energy nutrient. The unlikelihood of a bottle of sterile nutrient transforming itself into bacteria is therefore a manifestation not of the second law but rather of a putative new "slow growth" law that complexity, however defined, ought to obey: complexity ought not to increase quickly, except with low probability, but can increase slowly, for example, over geological time.

Computational Universality

Computational universality means the ability of a system to be pro-gramed through its initial conditions to simulate any digital computa-tion. While it is an eminently mathematical property, it is still too functional to be a good measure of complexity of physical states; it does not distinguish between a system capable of complex behavior and one in which the complex behavior has actually occurred. As a concrete ex-ample, it is known that classical billiard balls (Fredkin & Toffoli 1982), moving in a simple periodic potential, can be prepared in an initial condition to perform any computation; but if such a special initial con-dition has not been prepared, or if it has been prepared but the com-putation has not yet been performed, then the billiard-ball configuration does not deserve to be called complex. Much can be said about the theory of universal computers; here I note that their existence implies that the input-output relation of any one of them is a microcosm of all of deductive logic, and in particular of all axiomatizable physical theo-ries; moreover the existence of *efficiently* universal computers, which can simulate other computers with at most additive increase in pro-gram size and typically polynomial increase in execution time, allows the development of nearly machine-independent (and thus authorita-tive and absolute) theories of algorithmic information and computa-tional time/space complexity.

Computational Time/Space Complexity

Computational time/space complexity is the asymptotic difficulty (e.g., polynomial vs. exponential time in the length of its argument) of com-puting a function (Garey & Johnson 1979). By diagonal methods analo-gous to those used to show the existence of uncomputable functions, one can construct arbitrarily hard-to-compute computable functions. It is not immediately evident how a measure of the complexity of *func-tions* can be applied to *states* of physical models.

Algorithmic Information

Algorithmic information (also called algorithmic entropy or Solomonoff-Kolmogorov-Chaitin complexity) is the size in bits of the most con-cise universal computer program to generate the object in question (Ahlers & Walden 1980, Chaitin 1975, 1987, Levin 1984, Zurek 1989,

Zvonkin & Levin 1970). Algorithmic entropy is closely related to statistically defined entropy, the statistical entropy of an ensemble being, for any concisely describable ensemble, very nearly equal to the ensemble average of the algorithmic entropy of its members; but for this reason algorithmic entropy corresponds intuitively to randomness rather than to complexity. Just as the intuitively complex human body is intermediate in entropy between a crystal and a gas, so an intuitively complex genome or literary text is intermediate in algorithmic entropy between a random sequence and a prefectly orderly one.

Long-range Order

Long-range order, the existence of statistical correlations between arbitrarily remote parts of a body, is an unsatisfactory complexity measure, because it is present in such intuitively simple objects as perfect crystals.

Long-range Mutual Information

Long-range mutual information (remote nonadditive entropy) is the amount by which the joint entropy of two remote parts of a body exceeds the sum of their individual entropies. In a body with long-range order it measures the amount, rather than the range, of correlations. Remote mutual information arises for rather different reasons in equilibrium and nonequilibrium systems, and much more of it is typically present in the latter (Bennett 1987). In equilibrium systems, remote nonadditivity of the entropy is at most a few dozen bits and is associated with the order parameters, for example, magnetic or crystalline order in a solid. Correlations between remote parts of such a body are propagated via intervening portions of the body sharing the same value of the order parameter. By contrast, in nonequilibrium systems, much larger amounts of nonadditive entropy may be present, and the correlations need not be propagated via the intervening medium. Thus the contents of two newspaper dispensers in the same city is typically highly correlated, but this correlation is not mediated by the state of the intervening air (except for weather news). Rather it reflects each newspaper's descent from a common causal origin in the past. Similar correlations exist between genomes and organisms in the biosphere, reflecting the shared frozen accidents of evolution. This sort of long-range mutual information, not mediated by the intervening medium, is an attractive complexity measure in many respects, but it fails to obey the putative slow-growth law mentioned earlier: quite trivial processes of randomization and redistribution, for example, smashing a piece of glass and stirring up the pieces or replicating and stirring a batch of random meaningless DNA, generate enormous amounts of remote nonadditive entropy very quickly.

Logical Depth

Logical depth is the execution time required to generate the object in question by a nearly incompressible universal computer program, that is, one not itself computable as output of a significantly more concise program. Logical depth computerizes the Occam's razor paradigm, with programs representing hypotheses and outputs representing phenomena, and considers a hypothesis plausible only if it cannot be reduced to a simpler (more concise) hypothesis. Logically deep objects, in other words, contain internal evidence of having been the result of a long computation of slow-to-simulate dynamical process and could not plausibly have originated otherwise. Logical depth satisfies the slow-growth law by construction.

Thermodynamic Depth

The amount of entropy produced during a state's actual evolution has been proposed as a measure of complexity by Lloyd and Pagels (1988). Thermodynamic depth can be very system dependent: Some systems arrive at very trivial states through much dissipation, others at very nontrivial states with little dissipation.

Self-similar Structures and Chaotic Dynamics

Self-similar structures are striking to look at, and some intuitively complex entities are self-similar or at least hierachical in structure or function; but others are not. Moreover, some self-similar structures are rapidly computable, for example, by deterministic cellular automation rules. With regard to chaotic dynamics, Wolfram (1995) distinguished between "homoplectic" processes, which generate macroscopically random behavior by amplifying the noise in their initial and boundary conditions, and a more conjectural "autoplectic" type of process, which would generate macroscopically pseudoramdom behavior autonomously in the absence of noise and in the presence of noise would persist in reproducing the same pseudorandom sequence despite the noise. Such a noise-resistant process would have the possibility of evolving toward a deep state, containing internal evidence of a long history. A homoplectic processes, on the other hand, should produce only shallow states, containing evidence of that portion of the history recent enough not to have been swamped by dynamically amplified environmental noise.

Why: Usefulness of a Formal Measure of Complexity

Aside from their nonspecific usefulness in clarifying intuition, formal measures of complexity such as logical depth, as well as measures of randomness and correlation (e.g., algorithmic entropy and remote mutual information) raise a number of potentially decidable issues in statistical physics and the theory of computation.

Theory of Computation

The conjectured inequality of the complexity classes P and $PSPACE$ is a necessary condition, and the stronger conjecture of the existence of "one-way" functions (Bennett 1988; Levin 1985) is a sufficient condition for certain very idealized physical models (e.g., billiard balls) to generate logical depth efficiently.

Computationally Universal Model Systems

Which model systems in statistical mechanics are computationally universal? The billiard-ball model, consisting of hard spheres colliding with one another and with a periodic array of fixed mirrors in two dimensions, is computationally universal on a dynamically unstable set of trajectories measuring zero. In this model, the number of degrees of freedom is proportional to the space requirement of the computation, since each billiard ball encodes one bit. Probably the mirrors could be replaced by a periodic wind of additional balls, moving in the third dimension so that one "wind" ball crosses the plane of computation at time and location of each potential mirror collision and transfers the same momentum as the mirror would have done. This mirrorless model would have a number of degrees of freedom proportional to the time-space product of the computation being simulated. One might also ask whether a dynamical system with a fixed number of degrees of freedom, perhaps some version of the three-body problem, might be computationally universal. Such a model, if it exists, would not be expected to remain computationally universal in the presence of noise.

Error-correcting Computation

What collective phenomena suffice to allow error-correcting computation and the generation of complexity to proceed despite the locally destructive effects of noise? In particular, how does dissipation favor the generation and maintenance of complexity in noisy systems?

- Dissipation allows error-correction, a many-to-one mapping in phase space.
- Dissipative systems are exempt from the Gibbs phase rule. In typical d-dimensional equilibrium systems with short-ranged interactions, barring symmetries or accidental degeneracy of parameters such as occurs on a coexistence line, there is a unique thermodynamic phase of lowest free energy (Bennett & Grinstein 1985). This renders equilibrium systems ergodic and unable to store information reliably in the presence of "hostile" (i.e., symmmetry-breaking) noise. Analogous dissipative systems, because they have no defined free energy in d dimensions, are exempt from this rule. A $(d + 1)$-dimensional free energy can be defined, but varying the parameters of the d-dimensional model does not in general destabilize one phase relative to another.
- What other properties besides irreversibility does a system need to take advantage of the exemption from the Gibbs phase rule? In general the problem is to correct erroneous regions, in which the data or computation locally differs from that originally stored or programed into the system. These regions, which may be of any finite size, arise spontaneously due to noise and to subsequent propagation of errors through the system's normal dynamics. Local majority voting over a symmetric neighborhood, as in the Ising model at low temperature, is insufficient to suppress islands when the noise favors their growth. Instead of true stability, one has a metastable situation in which small islands are suppressed by surface tension, but large islands grow.

Two methods are known for achieving absolute stability in the presence of symmetry-breaking noise.

Anisotropic voting rules (Bennett & Grinstein 1985; Gacs & Reif 1985; Toom 1980) in two or more dimensions contrive to shrink arbitrarily large islands by differential motion of their boundaries. The rule is such that any island, while it may grow in some directions, shrinks in others; eventually the island becomes surrounded by shrinking facets only and disappears. The requisite anisotropy need not be present initially but may arise through spontaneous symmetry-breaking.

Hierarchical voting rules (Gacs 1986) are complex rules, in one or more dimensions, that correct errors by a programmed hierarchy of blockwise majority voting. The complexity arises from the need of the rule to maintain the hierarchical structure, which exists only in software.

Self-organization

Is "self-organization," the spontaneous increase of complexity, an asymptotically qualitative phenomenon like phase transitions? In other words, are there reasonable models whose complexity, starting from a simple uniform initial state, not only spontaneously increases but does so without bound in the limit of infinite space and time? Adopting logical depth as the criterion of complexity, this would mean that for arbitrarily large times t most parts of the system at time t would contain structures that could not plausibly have been generated in time much less than t. A positive answer to this question would not explain the history of our finite world but would suggest that its quantitative complexity can be legitimately viewed as an approximation to a well-defined property of infinite systems. On the other hand, a negative answer would suggest that our world should be compared to chemical reaction-diffusion systems that self-organize on a macroscopic but finite scale, or to hydrodynamic systems that self-organize on a scale determined by their boundary conditions, and that the observed complexity of our world may not be "spontaneous" but rather heavily conditioned by the anthropic requirement that it produce observers.

NOTE

Many of the ideas in this chapter were shaped in years of discussions with Gregory Chaitin, Rolf Landauer, Peter Gacs, Geoff Grindstein, and Joel Lebowitz.

REFERENCES

Ahlers, G., and R. W. Walden. 1980. "Turbulence near Onset of Convection." *Physical Review Letters* 44: 445.

Barrow, J. D., and F. J. Tipler. 1986. *The Anthropic Cosmological Principle.* Oxford: Oxford University Press.

Bennett, Charles H. 1982. "The Thermodynamics of Computation—A Review." *International Journal of Theoretical Physics* 21: 905–940.

Bennett, C. H. 1986. "On the Nature and Origin of Complexity in Discrete, Homogeneous, Locally-Interacting Systems." *Foundations of Physics* 16: 585–592.

Bennett, C. H. 1987. "Information, Dissipation, and the Definition of Organizations." In *Emerging Syntheses in Science*, edited by David Pines. Reading, Mass.: Addison-Wesley.

Bennett, C. H. 1988. "Logical Depth and Physical Complexity." In *The Universal Turing Machine: A Half-Century Survey*, edited by Rolf Herken. Oxford: Oxford University Press.

Bennett, C. H., & G. Grinstein. 1985. "On the Role of Irreversibility in Stabilizing Complex and Nonergodic Behavior in Locally Interacting Discrete Systems." *Physical Review Letters* 55: 657–660.

Chaitin, G. J. 1975. "A Theory of Program Size Formally Identical to Information Theory." *Journal of the Association of Computing Machinery* 22: 329–340.

Chaitin, G. J. 1987. *Algorithmic Information Theory.* Cambridge: Cambridge University Press.

Fredkin, E., and T. Toffoli. 1982. "Conservative Logics." *International Journal of Theoretical Physics* 21: 219–253.

Gacs, P. 1986. "Reliable Computation with Cellular Automata." *Journal of Computer and Systems Sciences* 32: 15–78.

Gacs, P., and J. Reif. 1985. "A Simple Three-Dimensional Real-Time Reliable Cellular Array." *Proceedings of the Seventeenth Association of Computing Machinery Symposium on the Theory of Computing*: 388–395.

Garey, M., and D. Johnson. 1979. *Computers and Intractability: A Guide to NP Completeness.* San Fransisco: Freeman.

Levin, L. A. 1984. "Randomness Conservation Inequalities: Information and Independence in Mathematical Theories." *Information and Control* 61: 15–37 (1980).

Levin, L. A. 1985. "One-Way Functions and Pseudorandom Generators." *Proceedings of the Seventeenth Association of Computing Machinery Symposium on Theory of Computing.*

Lloyd, L., and H. Pagels. 1988. "Complexity as Thermodynamic Depth." *Annals of Physics* 188: 186–213.

Toom, A. L. 1980. "Multicomponent Systems." In *Advances in Probability*, vol. 6, edited by R. L. Dobrushin. New York: Dekker.

Wolfram, S. 1995. "Origins of Randomness in Physical Systems." *Physical Review Letters* 55: 449–452.

Zurek, W. H. 1989. "Algorithmic Randomness and Physical Entropy." *Physical Review* A 40: 4731–4751.

Zvonkin, A. K., and L. A. Levin. 1970. "The Complexity of Finite Objects and the Development of the Concepts of Information and Randomness by Means of the Theory of Algorithms." *Russian Mathematical Surveys* 256: 83–124.

PART II

The Concept of Information in Physics and Biology

FOUR

THE EMERGENCE
OF AUTONOMOUS AGENTS

Stuart Kauffman

Lecturing in Dublin, one of the twentieth century's most famous physicists set the stage of contemporary biology during the war-heavy year of 1944. Given Erwin Schrödinger's towering reputation as the discoverer of the Schrödinger equation, the fundamental formulation of quantum mechanics, his public lectures and subsequent book were bound to draw high attention. But no one, not even Schrödinger himself, was likely to have foreseen the consequences. Schrödinger's *What Is Life?* is credited with inspiring a generation of physicists and biologists to seek the fundamental character of living systems. Schrödinger brought quantum mechanics, chemistry, and the still poorly formulated concept of "information" into biology. He is the progenitor of our understanding of DNA and the genetic code. Yet as brilliant as was Schrödinger's insight, I believe he missed the center.

In my previous two books, I laid out some of the growing reasons to think that evolution was even richer than Darwin supposed. Modern evolutionary theory, based on Darwin's concept of descent with heritable variations that are sifted by natural selection to retain the adaptive changes, has come to view selection as the sole source of order in biological organisms. But the snowflake's delicate sixfold symmetry tells us that order can arise without the benefit of natural selection. *Origins of Order* and *At Home in the Universe* give good grounds to think that much of the order in organisms, from the origin of life itself to the stunning order in the development of a newborn child from a fertilized egg, does not reflect selection alone. Instead, much of the order in organisms, I believe, is self-organized and spontaneous. Self-organization mingles with natural selection in barely understood ways to yield the magnificence of our teeming biosphere. We must, therefore, expand evolutionary theory.

Yet we need something far more important than a broadened evolutionary theory. Despite my valid insights in my own two books, and despite the fine work of many others, including the brilliance manifest in the past three decades of molecular biology, the core of life itself remains shrouded from view. We know chunks of molecular machinery, metabolic pathways, means of membrane biosynthesis—we know many of the parts and many of the processes. But what makes a cell alive is still not clear to us. The center is still mysterious.

What follows here is a brief introduction to the themes that have been explained more fully in my more recent book, *Investigations*.

My first efforts had begun with twin questions. First, in addition to the known laws of thermodynamics, could there possibly be a fourth law of thermodynamics for open thermodynamic systems, some law that governs biospheres anywhere in the cosmos or the cosmos itself? Second, living entities—bacteria, plants, and animals—manipulate the world on their own behalf: the bacterium swimming upstream in a glucose gradient that is easily said to be going to get "dinner"; the paramecium, cilia beating like a Roman warship's oars, hot after the bacterium; we humans earning our livings. Call the bacterium, paramecium, and us humans "autonomous agents," able to act on our own behalf in an environment.

My second and core question became, What must a physical system be to be an autonomous agent? Make no mistake, we autonomous agents mutually construct our biosphere even as we coevolve in it. Why and how this is so is a central subject of the following pages.

From the outset, there were, and remain, reasons for deep skepticism about this enterprise. First, there are very strong arguments to say that there can be no general law for open thermodynamic systems. The core argument is simple to state. Any computer program is an algorithm that, given data, produces some sequence of output, finite or infinite. Computer programs can always be written in the form of a binary symbol string of 1 and 0 symbols. All possible binary symbol strings are possible computer programs. Hence there is a countable, or denumerable, infinity of computer programs. A theorem states that for most computer programs, there is no compact description of the printout of the program. Rather, we must just unleash the program and watch it print what it prints. In short, there is no shorter description of the output of the program than that which can be obtained by running the program itself. If by the concept of a "law" we mean a compact description, ahead of time, of what the computer program will print, then for any such program, there can be no law that allows us to predict what the program will actually do ahead of the actual running of the program.

The next step is simple. Any such program can be realized on a universal Turing machine such as the familiar computer. But that computer is an open nonequilibrium thermodynamic system, its openness visibly realized by the plug and power line that connects the computer to the electric power grid. Therefore, and I think this conclusion is cogent, there can be no general law for all possible nonequilibrium thermodynamic systems.

So why was I conjuring the possibility of a general law for open thermodynamic systems? Clearly, no such general law can hold for all open thermodynamic systems.

But hold a moment. It is we humans who conceived and built the intricate assembly of chips and logic gates that constitute a computer, typically we humans who program it, and we humans who contrived the entire power grid that supplies the electric power to run the computer itself. This assemblage of late-twentieth-century technology did not assemble itself. We built it.

On the other hand, no one designed and built the biosphere. The biosphere got itself constructed by the emergence and persistent co-evolution of autonomous agents. If there cannot be general laws for all open thermodynamic systems, might there be general laws for thermo-dynamically open but self-constructing systems such as biospheres? I believe that the answer is yes. Indeed, among those candidate laws to be discussed in this book is a candidate fourth law of thermodynamics for such self-constructing systems.

To roughly state the candidate law, I suspect that biospheres maximize the average secular construction of the diversity of autonomous agents and the ways those agents can make a living to propagate further. In other words, on average, biospheres persistently increase the diversity of what can happen next. In effect, as we shall see later, biospheres may maximize the average sustained growth of their own "dimensionality."

Thus, I soon began to center on the character of the autonomous agents whose coevolution constructs a biosphere. I was gradually led to a labyrinth of issues concerning the core features of autonomous agents able to manipulate the world on their own behalf. It may be that those core features capture a proper definition of life and that definition differs from the one Schrödinger found.

To state my hypothesis abruptly and without preamble, I think an autonomous agent is a self-reproducing system able to perform at least one thermodynamic work cycle.

Following an effort to understand what an autonomous agent might be—which, as just noted, involves the concept of work cycles—I was led to the concepts of work itself, constraints, and work as the constrained release of energy. In turn, this led to the fact that work itself is often used to construct constraints on the release of energy that then constitutes further work. So we confront a virtuous cycle: Work constructs constraints, yet constraints on the release of energy are required for work to be done. Here is the heart of a new concept of "organization" that is not covered by our concepts of matter alone, energy alone, entropy alone, or information alone. In turn, this led me to wonder about the relation between the emergence of constraints in the universe and in a biosphere, and the diversification of patterns of the constrained release of energy that alone constitute work and the use of that work to build still further constraints on the release of energy. How do biospheres construct themselves or how does the universe construct itself?

These considerations led to the role of Maxwell's demon, one of the major places in physics where matter, energy, work, and information come together. The central point of the demon is that by making measurements on a system, the information gained can be used to extract work. I made a new distinction between measurements the demon might make that reveal features of nonequilibrium systems that can be used to extract work and measurements he might make of the nonequiibrium system that cannot be used to extract work. How does the demon know what features to measure? And, in turn, how does work actually come to be extracted by devices that measure and detect displacements from equilibrium from which work can, in principle, be obtained? An example of such a device is a windmill pivoting to face the wind and then extracting work by the wind turning its vanes. Other examples are the rhodopsin molecule of a bacterium responding to a photon of light or a chloroplast using the constrained release of the energy of light to construct high-energy sugar molecules. How do such devices come into existence in the unfolding universe and in our biosphere? How does the vast web of constraint construction and constrained energy release used to construct yet more constraints happen into existence in the biosphere? In the universe itself? The answers appear not to be present in contemporary physics, chemistry, or biology. But a coevolving biosphere accomplishes just this coconstruction of propagating organization.

Thus, in due course, I struggled with the concept of organization itself, concluding that our concepts of entropy and its negative, Shannon's information theory (which was developed initially to quantify telephonic traffic and had been greatly extended since then) entirely miss the central issues. What is happening in a biosphere is that autonomous agents are coconstructing and propagating organizations of work, of constraint construction, and of task completion that continue to propagate and proliferate diversifying organization.

This statement is just plain true. Look out your window, burrow down a foot or so, and try to establish what all the microscopic life is busy doing and building and has done for billions of years, let alone the macroscopic ecosystem of plants, herbivores, and carnivores that is slipping, sliding, hiding, hunting, and bursting with flowers and leaves outside your window. So, I think, we lack a concept of propagating organization.

Then too there is the mystery of the emergence of novel functionalities in evolution where none existed before: hearing, sight, flight, language. Whence this novelty? I was led to doubt that we could prestate the novelty. I came to doubt that we could finitely prestate all possible adaptations that might arise in a biosphere. In turn, I was led to doubt that we can prestate the "configuration space" of a biosphere.

But how strange a conclusion. In statistical mechanics, with its famous liter box of gas as an isolated thermodynamic system, we can prestate the configuration space of all possible positions and momenta of the gas particles in the box. Then Ludwig Boltzmann and Willard Gibbs taught us how to calculate macroscopic properties such as pressure and temperature as equilibrium averages over the configuration space. State the laws and the initial and boundary conditions, then calculate; Newton taught us how to do science this way. What if we cannot prestate the configuration space of a biosphere and calculate with Newton's "method of fluxions," the calculus, from initial and boundary conditions and laws? Whether we can calculate or not does not slow down the persistent evolution of novelty in the biosphere. But a biosphere is just another physical system. So what in the world is going on? Literally, what in the world is going on?

We have much to investigate. At the end, I think we will know more than at the outset.

It is well to return to Schrödinger's brilliant insights and his attempt at a central definition of life as a well-grounded starting place.

Schrödinger's *What Is Life?* provided a surprising answer to his enquiry about the central character of life by posing a core question: What is the source of the astonishing order in organisms? The standard—and, Schrödinger argued, incorrect—answer lay in statistical physics. If an ink drop is placed in still water in a petri dish, it will diffuse to a uniform equilibrium distribution. That uniform distribution is an average over an enormous number of atoms or molecules and is not due to the behavior of individual molecules. Any local fluctuations in ink concentration soon dissipate back to equilibrium.

Could statistical averaging be the source of order in organisms? Schrödinger based his argument on the emerging field of experimental genetics and the recent data on X-ray induction of heritable genetic mutations. Calculating the "target size" of such mutations, Schrödinger realized that a gene could comprise at most a few hundred or thousand atoms.

The sizes of statistical fluctuations familiar from statistical physics scale as the square root of the number of particles, N. Consider tossing a fair coin 10,000 times. The result will be about 50 percent heads, 50 percent tails, with a fluctuation of about 100, which is the square root of 10,000. Thus, a typical fluctuation from 50:50 heads and tails is 100/10,000, or 1 percent. Let the number of coin flips be 100 million, then the fluctuations are its square root, or 10,000. Dividing, 10,000/100,000,000 yields a typical deviation of .01 percent from 50:50.

Schrödinger reached the correct conclusion: If genes are constituted by as few as several hundred atoms, the familiar statistical fluctuations predicted by statistical mechanics would be so large that heritability would be essentially impossible. Spontaneous mutations would happen at a frequency vastly larger than observed. The source of order must lie elsewhere.

Quantum mechanics, argued Schrödinger, comes to the rescue of life. Quantum mechanics ensures that solids have rigidly ordered molecular structures. A crystal is the simplest case. But crystals are structurally dull. The atoms are arranged in a regular lattice in three dimensions. If you know the positions of all the atoms in a minimal-unit crystal, you know where all the other atoms are in the entire crystal. This overstates the case, for there can be complex defects, but the point is clear. Crystals have very regular structures, so the different parts of the crystal, in some sense, all "say" the same thing. As shown hereafter, Schrödinger translated the idea of "saying" into the idea of "encoding." With that leap, a regular crystal cannot encode much "information' All the information is contained in the unit cell.

If solids have the order required but periodic solids such as crystals are too regular, then Schrödinger puts his bet on aperiodic solids. The stuff of the gene, he bets, is some form of aperiodic crystal. The form of the aperiodicity will contain some kind of microscopic code that somehow controls the development of the organism. The quantum character of the aperiodic solid will mean that small discrete changes, or mutations, will occur. Natural selection, operating on these small discrete changes, will select out favorable mutations, as Darwin hoped.

Fifty years later, I find Schrödinger's argument fascinating and brilliant. At once he envisioned what became, by 1953, the elucidation of the structure of DNA's aperiodic double helix by James Watson and Francis Crick, with the famously understated comment in their original article that its structure suggests its mode of replication and its mode of encoding genetic information.

Fifty years later we know very much more. We know the human genome harbors some 80,000 to 100,000 "structural genes," each encoding the RNA that, after being transcribed from the DNA, is translated according to the genetic code to a linear sequence of amino acids, thereby constituting a protein. From Schrödinger to the establishment of the code required only about 20 years.

Beyond the brilliance of the core of molecular genetics, we understand much concerning developmental biology. Humans have about 260 different cell types: liver, nerve, muscle. Each is a different pattern of expression of the 80,000 or 100,000 genes. Since the work of François Jacob and Jacques Monod 35 years ago, biologists have understood that the protein transcribed from one gene might turn other genes on or off. Some vast network of regulatory interactions among genes and their products provides the mechanism that marshals the genome into the dance of development.

We have come close to Schrödinger's dream. But have we come close to answering his question, What is life? The answer almost surely is no. I am unable to say, all at once, why I believe this, but I can begin to hint at an explanation. Life is doing something far richer than we may have dreamed, literally something incalculable. What is the place of law if, as hinted earlier, variables and configuration space cannot be prespecified for a biosphere, or perhaps a universe? Yet, I think, there are laws. And if these musings be true, we must rethink science itself.

Perhaps I can point again at the outset to the central question of an autonomous agent. Consider a bacterium swimming upstream in a glucose gradient, its flagellar motor rotating. If we naively ask "What is it doing?" we unhesitatingly answer something like "It's going to get dinner." That is, without attributing consciousness or conscious purpose, we view the bacterium as acting on its own behalf in an environment. The bacterium is swimming upstream in order to obtain the glucose it needs. Presumably we have in mind something like the Darwinian criteria to unpack the phrase "on its own behalf." Bacteria that do obtain glucose or its equivalent may survive with higher probability than those incapable of the flagellar motor trick, hence be selected by natural selection.

An autonomous agent is a physical system, such as a bacterium, that can act on its own behalf in an environment. All free-living cells and organisms are clearly autonomous agents. The quite familiar, utterly astonishing feature of autonomous agents—*E. coli*, paramecia, yeast cells, algae, sponges, flat worms, annelids, all of us—is that we do, everyday, manipulate the universe around us. We swim, scramble, twist, build, hide, snuffle, pounce.

Yet the bacterium, the yeast cell, and we all are just physical systems. Physicists, biologists, and philosophers no longer look for a mysterious élan vital, some ethereal vital force that animates matter. Which leads immediately to the central, and confusing, question: What must a physical system be such that it can act on its own behalf in an environment? What must a physical system be such that it constitutes an autonomous agent? I will leap ahead to give now my tentative answer: A molecular autonomous agent is a self-reproducing molecular system able to carry out one or more thermodynamic work cycles.

All free-living cells are, by this definition, autonomous agents. To take a simple example, our bacterium with its flagellar motor rotating and swimming upstream for dinner is, in point of plain fact, a self-reproducing molecular system that is carrying out one or more thermodynamic work cycles. So is the paramecium chasing the bacterium, hoping for its own dinner. So is the dinoflagellate hunting the paramecium sneaking up on the bacterium. So are the flower and flatworm. So are you and I.

It will take a while to fully explore this definition. Unpacking its implications reveals much that I did not remotely anticipate. An early insight is that an autonomous agent must be displaced from thermodynamic equilibrium. Work cycles cannot occur at equilibrium. Thus, the concept of an agent is, inherently, a nonequilibrium concept. So too at the outset it is clear that this new concept of an autonomous agent is not contained in Schrödinger's answer. Schrödinger's brilliant leap to aperiodic solids encoding the organism that unleashed mid-twentieth-century biology appears to be but a glimmer of a far larger story.

Footprints of Destiny:
The Birth of Astrobiology

The telltale beginnings of that larger story are beginning to be formulated. The U.S. National Aeronautics and Space Agency has had a long program in "exobiology," the search for life elsewhere in the universe. Among its well-known interests are SETI, a search for extraterrestrial life, and the Mars probes. Over the past three decades, a sustained effort has included a wealth of experiments aiming at discovering the abiotic origins of the organic molecules that are the building blocks of known living systems.

In the summer of 1997, NASA was busy attempting to formulate what it came to call "astrobiology," an attempt to understand the origin, evolution, and characteristics of life anywhere in the universe. Astrobiology does not yet exist—it is a field in the birthing process. Whatever the area comes to be called as it matures, it seems likely to be a field of spectacular success and deep importance in the coming century. A hint of the potential impact of astrobiology came in August 1997 with the tentative but excited reports of a Martian meteorite found in Antarctica that, NASA scientists announced, might have evidence of early Martian microbial life. The White House organized the one-day "Space Conference," to which I was pleased to be invited. Perhaps 35 scientists and scholars gathered in the Old Executive Office Building for a meeting. led by Vice President Gore. The vice president began the meeting with a rather unexpected question to the group: If it should prove true that the Martian rock actually harbored fossilized microbial life, what would be the least interesting result?

The room was silent, for a moment. Then Stephen Jay Gould gave the answer many of us must have been considering: "Martian life turns out to be essentially identical to Earth life, same DNA, RNA, proteins, code." Were it so, then we would all envision life flitting from planet to planet in our solar system. It turns out that a minimum transit time for a fleck of Martian soil kicked into space to make it to Earth is about 15,000 years. Spores can survive that long under desiccating conditions.

"And what," continued the vice president, "would be the most interesting result?" Ah, said many of us, in different voices around the room: Martian life is radically different from Earth life.

If radically different, then . . .

If radically different, then life must not be improbable.

If radically different, then life may be abundant among the myriad stars and solar systems, on far planets hinted at by our current astronomy.

If radically different and abundant, then we are not alone.

If radically different and abundant, then we inhabit a universe rife with the creativity to create life.

If radically different, then—thought I of my just-published second book—we are at home in the universe.

If radically different, then we are on the threshold of a new biology, a "general biology" freed from the confines of our known example of Earth life.

If radically different, then a new science seeking the origins, evolution, characteristics, and laws that may govern biospheres anywhere.

A general biology awaits us. Call it astrobiology if you wish. We confront the vast new task of understanding what properties and laws, if any, may characterize biospheres anywhere in the universe. I find the prospect stunning. I will argue that the concept of an autonomous agent will be central to the enterprise of a general biology.

A personally delightful moment arose during that meeting. The vice president, it appeared, had read *At Home in the Universe*, or parts of it. In *At Home*, and also in *Investigations*, I explore a theory I believe has deep merit, one that asserts that, in complex chemical reaction systems, self-reproducing molecular systems form with high probability.

The vice president looked across the table at me and asked, "Dr. Kauffman, don't you have a theory that in complex chemical reaction systems life arises more or less spontaneously?"

"Yes."

"Well, isn't that just sensible?"

I was, of course, rather thrilled, but somewhat embarrassed. "The theory has been tested computationally, but there are no molecular experiments to support it," I answered.

"But isn't it just sensible?" the vice president persisted.

I couldn't help my response, "Mr. Vice President, I have waited a long time for such confirmation. With your permission, sir, I will use it to bludgeon my enemies."

I'm glad to say there was warm laughter around the table. Would that scientific proof were so easily obtained. Much remains to be done to test my theory.

Many of us, including Mr. Gore, while maintaining skepticism about the Mars rock itself, spoke at that meeting about the spiritual impact of the discovery of life elsewhere in the universe. The general consensus was that such a discovery, linked to the sense of membership in a creative universe, would alter how we see ourselves and our place under all, all the suns. I find it a gentle, thrilling, quiet, and transforming vision.

Molecular Diversity

We are surprisingly well poised to begin an investigation of a general biology for such a study will surely involve the understanding of the collective behaviors of very complex chemical reaction networks. After all, all known life on earth is based on the complex webs of chemical reactions—DNA, RNA, proteins, metabolism, linked cycles of construction and destruction—that form the life cycles of cells. In the past decade we have crossed a threshold that will rival the computer revolution. We have learned to construct enormously diverse "libraries" of different DNA, RNA, proteins, and other organic molecules. Armed with such high-diversity libraries, we are in a position to begin to study the properties of complex chemical reaction networks.

To begin to understand the molecular diversity revolution, consider a crude estimate of the total organic molecular diversity of the biosphere. There are perhaps a hundred million species. Humans have about a hundred thousand structural genes, encoding that many different proteins. If all the genes within a species were identical, and all the genes in different species were at least slightly different, the biosphere would harbor about 10 trillion different proteins. Within a few orders of magnitude, 10 trillion will serve as an estimate of the organic molecular diversity of the natural biosphere. But the current technology of molecular diversity that generates libraries of more or less random DNA, RNA, or proteins now routinely produces a diversity of a hundred trillion molecular species in a single test tube.

In our hubris, we rival the biosphere.

The field of molecular diversity was born to help solve the problem of drug discovery. The core concept is simple. Consider a human hormone such as estrogen. Estrogen acts by binding to a specific receptor

protein; think of the estrogen as a "key" and the receptor as a "lock."
Now generate 64 million different small proteins, called peptides, say,
six amino acids in length. (Since there are 20 types of amino acids, the
number of possible hexamers is 20^6, hence 64 million.) The 64 million
hexamer peptides are candidate second keys, any one of which might
be able to fit into the same estrogen receptor lock into which estrogen
fits. If so, any such second key may be similar to the first key, estrogen,
and hence is a candidate drug to mimic or modulate estrogen.

To find such an estrogen mimic, take many identical copies of the
estrogen receptor, afix them to the bottom of a petri plate, and expose
them simultaneously to all 64 million hexamers. Wash off all the peptides that do not stick to the estrogen receptor and then recover those
hexamers that do stick to the estrogen receptor. Any such peptide is a
second key that binds the estrogen receptor locks and hence is a candidate estrogen mimic.

The procedure works, and works brilliantly. In 1990, George Smith
at the University of Missouri used a specific kind of virus, a filamen-
tous phage that infects bacteria. The phage is a strand of RNA that
encodes proteins. Among these proteins is the coat protein that pack-
ages the head of the phage as part of an infective phage particle. George
cloned random DNA sequences encoding random hexamer peptides
into one end of the phage coat protein gene. Each phage then carried
a different, random DNA sequence in its coat protein gene, hence made
a coat protein with a random six-amino-acid sequence at one end. The
initial resulting "phage display" libraries had about 20 million of the
64 million different possible hexamer peptides.

Rather than using the estrogen receptor and seeking a peptide estrogen
mimic that binds the estrogen receptor, George Smith used a monoclonal
antibody molecule as the analogue of the receptor and sought a hexamer
peptide that could bind the monoclonal antibody. Monoclonal antibody
technology allows the generation of a large number of identical antibody
molecules, hence George could use these as identical mock receptors.
George found that, among the 20 million different phage, about one in a
million would stick to his specific monoclonal antibody molecules. In fact,
George found 19 different hexamers binding to his monoclonal antibody.
Moreover, the 19 different hexamers differed from one another, on average, in three of the six amino acid positions. All had high affinity for his
monoclonal antibody target.

These results have been of very deep importance. Phage display is
now a central part of drug discovery in many pharmaceutical and bio-
technology companies. The discovery of "drug leads" is being trans-

formed from a difficult to a routine task. Work is being pursued not only using peptides but also using RNA and DNA sequences. Molecular diversity has now spread to the generation of high-diversity libraries of small organic molecules, an approach called "combinatorial chemistry." The promise is of high medical importance. As we understand better the genetic diversity of the human population, we can hope to create well-crafted molecules with increased efficacy as drugs, vaccines, enzymes, and novel molecular structures. When the capacity to craft such molecules is married, as it will be in the coming decades, to increased understanding of the genetic and cellular signaling pathways by which ontogeny is controlled, we will enter an era of "postgenomic" medicine. By learning to control gene regulation and cell signaling, we will begin to control cell proliferation, cell differentiation, and tissue regeneration to treat pathologies such as cancer, autoimmune diseases, and degenerative diseases.

But George Smith's experiments are also of immediate interest, and in surprising ways.

George's experiments have begun to verify the concept of a "shape space" put forth by George Oster and Alan Perelson of the University of California, Berkeley, and Los Alamos National Laboratory more than a decade earlier, In turn, shape space suggests "catalytic task space." Both are needed for an understanding of autonomous agents.

Oster and Perelson had been concerned about accounting for the fact that humans can make about a hundred million different antibody molecules. Why, they wondered. They conceived of an abstract shape space with perhaps seven or eight "dimensions." Three of these dimensions would correspond to the three spatial dimensions, length, height, and width, of a molecular binding site. Other dimensions might correspond to physical properties of the binding sites of molecules, such as charge, dipole moment, and hydrophobicity.

A point in shape space would represent a molecular shape. An antibody binds its shape complement, key and lock. But the precision with which an antibody can recognize its shape complement is finite. Some jiggle room is allowed. So an antibody molecule "covers" a kind of "ball" of complementary shapes in shape space. And then comes the sweet argument. If an antibody covers a ball, an actual volume, in shape space, then a finite number of balls will suffice to cover all of shape space. A reasonable analogy is that a finite number of ping-pong balls will fill up a bedroom.

But how big of a ping-pong ball in shape space is covered by one antibody? Oster and Perelson reasoned that in order for an immune system to protect an organism against disease, its antibody repertoire should cover a reasonable fraction of shape space. Newts, with about 10,000 different antibody molecules, have the minimal known antibody diversity. Perelson and Oster guessed that the newt repertoire must cover a substantial fraction, say about $1/e$—where e is the natural base for logarithms—or 37 percent of shape space. Dividing 37 percent by 10,000 gives the fractional volume of shape space covered by one antibody molecule. It follows that 100 million such balls, thrown at random into shape space and allowed to overlap one another, will saturate shape space. So, 100 million antibody molecules is all we need to recognize virtually any shape of the size scale of molecular binding sites.

And therefore the concept of shape space carries surprising implications. Not surprisingly, similar molecules can have similar shapes. More surprisingly, very different molecules can have the same shape. Examples include endorphin and morphine. Endorphin is a peptide hormone. When endorphin binds the endorphin brain receptor, a euphoric state is induced. Morphine, a completely different kind of organic molecule, binds the endorphin receptor as well, with well-known consequences. Still more surprising, a finite number of different molecules, about a hundred million, can constitute a universal shape library. Thus, while there are vastly many different proteins, the number of effectively different shapes may only be on the order of a hundred million.

If one molecule binding to a second molecule can be thought of as carrying out a "binding task," then about a hundred million different molecules may constitute a universal toolbox for all molecular binding tasks. So if we can now create libraries with 100 trillion different proteins, a millionfold in excess of the universal library, we are in a position to begin to study molecular binding in earnest.

But there may also be a universal enzymatic toolbox. Enzymes catalyze, or speed up, chemical reactions. Consider a substrate molecule undergoing a reaction to a product molecule. Physical chemists think of the substrate and product molecules as lying in two potential "energy wells," like a ball at the bottom of one of two adjacent bowls. A chemical reaction requires "lifting" the substrate energetically to the top of the barrier between the bowls. Physically, the substrate's bonds are maximally strained and deformed at the top of this potential barrier. The deformed molecule is called the "transition state." According to transition state theory, an enzyme works by binding to and stabilizing the transition state molecule, thereby lowering the potential barrier of the reaction. Since the probability that a molecule acquires enough energy to hop to the top of the potential barrier is exponentially less as the barrier height increases, the stabilization of the transition state by the enzyme can speed up the reaction by many orders of magnitude.

Think of a catalytic task space, in which a point represents a catalytic task, where a catalytic task is the binding of a transition state of a reaction. Just as similar molecules can have similar shapes, so too can similar reactions have similar transition states, hence such reactions constitute similar catalytic tasks. Just as different molecules can have the same shape, so too can different reactions have similar transition states, hence constitute the "same" catalytic task. Just as an antibody can bind to and cover a ball of similar shapes, an enzyme can bind to and cover a ball of similar catalytic tasks. Just as a finite number of balls can cover shape space, a finite number of balls can cover catalytic task space.

In short, a universal enzymatic toolbox is possible. Clues that such a toolbox is experimentally feasible come from many recent developments, including the discovery that antibody molecules, evolved to bind molecular features called epitopes, can actually act as catalysts.

Catalytic antibodies are obtained exactly as one might expect, given the concept of a catalytic task space. One would like an antibody molecule that binds the transition state of a reaction. But transition states are ephemeral. Since they last only fractions of a second, one cannot immunize with a transition state itself. Instead, one immunizes with a stable analogue of the transition shape; that is, one immunizes with a second molecule that represents the "same" catalytic task as does the transition state itself. Antibody molecules binding to this transition state analogue are tested. Typically, about 1 in 10 antibody molecules can function as at least a weak catalyst for the corresponding reaction.

These results even allow a crude estimate of the probability that a randomly chosen antibody molecule will catalyze a randomly chosen reaction. About one antibody in a hundred thousand can bind a randomly chosen epitope. About 1 in 10 antibodies that bind the transition state analogue act as catalysts. By this crude calculation, about one in a million antibody molecules can catalyze a given reaction.

This rough calculation is probably too high by several orders of magnitude, even for antibody molecules. Recent experiments begin to address the probability that a randomly chosen peptide or DNA or RNA sequence will catalyze a randomly chosen reaction. The answer for DNA or RNA appears to be about one in a billion to one in a trillion. If we now make libraries of a hundred trillion random DNA, RNA, and protein molecules, we may already have in hand universal enzymatic toolboxes. Virtually any reaction, on the proper molecular scale of reasonable substrates and products, probably has one or more catalysts in such a universal toolbox.

In short, among the radical implications of molecular diversity is that we already possess hundreds of millions of different molecular functions—binding, catalytic, structural, and otherwise.

In our hubris, we rival the biosphere.

In our humility, we can begin to formulate a general biology and begin to investigate the collective behaviors of hugely diverse molecular libraries. Among these collective behaviors must be life itself.

Life as an Emergent Collective Behavior
of Complex Chemical Networks

In the summer of 1996, Philip Anderson, a Nobel laureate in physics, and I accompanied Dr. Henry MacDonald, incoming director of NASA at Ames, Iowa, to NASA headquarters. Our purpose was to discuss a new linked experimental and theoretical approach to the origin-of-life problem with NASA administrator Dan Golden and his colleague, Dr. Wesley Huntress. I was excited and delighted.

As long ago as 1971, I had published my own first foray into the origin-of-life problem as a young assistant professor in the Department of Theoretical Biology at the University of Chicago. I had wondered if life must be based on template replicating nucleic acids such as DNA or RNA double helices and found myself doubting that standard assumption. Life, at its core, depends on autocatalysis, that is, reproduction. Most catalysis in cells is carried out by protein enzymes. Might there be general laws supporting the possibility that systems of catalytic polymers such as proteins might be self-reproducing? Proteins are, as noted, linear sequences of 20 kinds of standard amino acids. Consider, then, a first copy of a protein that has the capacity to catalyze a reaction by which two fragments of a potential second copy of that same protein might be ligated to make the second copy of the whole protein. Such a protein, A, say, 32 amino acids long, might act on two fragments, say, 15 amino acids and 17 amino acids in length, and ligate the two to make a second copy of the 32-amino-acid sequence.

But if one could imagine a molecule, A, catalyzing its own formation from its own fragments, could one not imagine two proteins, A and B, having the property that A catalyzes the formation of B by ligating *B*'s fragments into a second copy of *B*, while *B* catalyzes the formation of *A* by catalyzing the ligation of *A*'s fragments into a second copy of *A*? Such a little reaction system would be *collectively autocatalytic*. Neither *A* alone nor *B* alone would catalyze its own formation. Rather the *AB* system would jointly catalyze its reproduction from *A* and *B* fragments. But if *A* and *B* might achieve collective autocatalysis, might one envision a system with tens or hundreds of proteins, or peptides, that were collectively autocatalytic?

Might collective autocatalysis of proteins or similar polymers be the basic source of self-reproduction in molecular systems? Or must life be based on template replication, as envisioned by Watson and Crick, or as envisioned even earlier by Schrödinger in his aperiodic solid with its microcode? In view of the potential for a general biology, what, in fact, are the alternative bases for self-reproducing molecular systems here and anywhere in the cosmos? Which of these alternatives is more probable, here and anywhere?

By 1971 I had asked and found a preliminary answer to the following question: In a complex mixture of different proteins, where the proteins might be able to serve as candidates to ligate one another into still larger amino acid sequences, what are the chances that such a system will contain one or more collectively autocatalytic sets of molecules? The best current guess is that, as the molecular diversity of a reaction system increases, a critical threshold is reached at which collectively autocatalytic, self-reproducing chemical reaction networks emerge spontaneously.

If this view is correct, and the kinetic conditions for rapid reactions can be sustained, perhaps by enclosure of such a reproducing system in a bounding membrane vesicle, also synthesized by the system, the emergence of self-reproducing molecular systems may be highly probable. No small conclusion this: Life abundant, emergent, expected. Life spattered across megaparsecs, galaxies, galactic clusters. We as members of a creative, mysteriously unfolding universe. Moreover, the hypothesis is richly testable and is now in the early stages of testing.

One way or another, we will discover a second life—crouched under a Mars rock, frozen in time; limpid in some pool on Titan, in some test tube in Nebraska in the next few decades. We will discover a second life, one way or another.

What monumental transformations await us, proudly postmodern, mingled with peoples on this very globe still wedded to archetypes thousands of years old.

The Strange Thing about the
Theory of Evolution

We do not understand evolution. We live it with moss, fruit, fin, and quill fellows. We see it since Darwin. We have insights of forms and their formation, won from efforts since Aristotle codified the embryological investigations that over 25 centuries ago began with the study of deformed fetuses in sacrificial animals.

But we do not understand evolution.

"The strange thing about the theory of evolution," said one of the Huxleys (although I cannot find which one), "is that everyone thinks he understands it." How very well put in that British fashion Americans can admire but not emulate. ("Two peoples separated by a common language," as Churchill dryly put it.)

The strange thing about the theory of evolution is that everyone thinks he understands it. How very true. It seems, of course, so simple. Finches hop around the Galapagos, occasionally migrating from island to island. Small and large beaks serve for different seeds. Beaks fitting seeds feed the young. Well-wrought beaks are selected. Mutations are the feedstock of heritable variation in a population. Populations evolve by mutation, mating, recombination, and selection to give the well-marked varieties that are, for Darwin, new species. Phylogenies bushy in the biosphere. "We're here, we're here," cry all for their typical four-million-year-stay along the four-billion-year pageant.

"We're here!"

But how?

How, in many senses. First, Darwin's theory of evolution is a theory of descent with modification. It does not yet explain the genesis of forms, but the trimmings of the forms once they are generated. "Rather like achieving an apple tree by trimming off all the branches," said a late-nineteenth-century skeptic.

How, in the most fundamental sense: Whence life in the first place? Darwin starts with life already here. Whence life is the stuff of all later questions about whence the forms to sift.

How, in still a different sense. Darwin assumed gradualism. Most variation would be minor. Selection would sift these insensible alterations, a bit more lift, a little less drag, until the wing flew faultless in the high-hoped sky, a falcon's knot-winged, claw-latching dive to dine.

But whence the gradualism itself? It is not God given, but true, that organisms are hardly affected by most mutations. Most mutations do have little effect, some have major effects. In *Drosophila*, many mutants make small modifications in bristle number, color, shape. A few change wings to legs, eyes to antennae, heads to genitalia. Suppose that all mutations were of dramatic effect. Suppose, to take the limiting philosophical case, that all mutations were what geneticists call "lethals." Since, indeed, some mutations are lethals, one can, a priori, imagine creatures in which all mutations were lethal prior to having offspring. Might be fine creatures, too, in the absence of any mutations, these evolutionary descendants of, well, of what? And progenitors of whom? No pathway to or from these luckless ones.

Thus, evolution must somehow be crafting the very capacity of creatures to evolve. Evolution nurtures herself! But not yet in Darwin's theory, nor yet in ours.

Take another case—sex. Yes, it captures our attention, and the attention of most members of most species. Most species are sexual. But why bother? Asexuals, budding quietly wherever they bud, require only a single parent. We plumaged ones require two, a twofold loss in fitness.

Why sex? The typical answer, to which I adhere, is that sexual mating gives the opportunity for genetic recombination. In genetic recombination, the double chromosome complement sets, maternal and paternal homologues, pair up, break, and recombine to yield offspring chromosomes the left half of which derives from one parental chromosome, the right half of which derives from the other parental chromosome.

Recombination is said to be a useful "search procedure" in an evolving population. Consider, a geneticist would say, two genes, each with two versions, or alleles: A and a for the first gene, B and b for the second gene. Suppose A confers a selective advantage compared to a, and B confers an advantage with respect to b. In the absence of sex, mating, and recombination, a rabbit with A and b would have to wait for a mutation to convert b to B. That might take a long time. But with mating and recombination, a rabbit with A on the left end of a maternal chromosome and B on the right end of the homologous paternal chromosome might experience recombination. A and B would now be on a single chromosome, hence be passed on to the offspring. Recombination, therefore, can be a lot faster than waiting for mutation to assemble the good, AB chromosome.

But it is not so obvious that recombination is a good idea after all. At the molecular level, the recombination procedure is rather like taking an airplane and a motorcycle, breaking both in half, and using spare bolts to attach the back half of the airplane to the front half of the motorcycle. The resulting contraption seems useless for any purpose.

In short, the very usefulness of recombination depends on the gradualness that Darwin assumed. The basic idea is simple. Consider a set of all possible frogs, each with a different genotype. Locate each frog in a high-dimensional "genotype space," each next to all genotypes that differ from it by a single mutation. Imagine that you can measure the fitness of each frog. Graph the fitness of each frog as a height above that position in genotype space. The resulting heights form a *fitness landscape* over the genotype space, much as the Alps form a mountainous landscape over part of Europe.

In the fitness landscape, image, mutation, recombination, and selection can conspire to pull evolving populations upward toward the peaks of high fitness. But not always. It is relatively easy to show that recombination is only a useful search procedure on smooth fitness landscapes. The smoothness of a fitness landscape can be defined mathematically by a correlation function giving the similarity of fitnesses, or heights, at two points on the landscape separated by a mutational distance. In the Alps, most nearby points are of similar heights, except for cliffs, but points 50 kilometers apart can be of very different heights. Fifty kilometers is beyond the correlation length of the Alps.

There is good evidence that recombination is only a useful search strategy on smooth, highly correlated landscapes, where the high peaks all cluster near one another. Recombination, half airplane–half motorcycle, is a means to look "between" two positions in a high-dimensional space. Then if both points are in the region of high peaks, looking between those two points is likely to uncover further new points of high fitness, or points on the slopes of even higher peaks. Thereafter, further mutation, recombination, and selection can bring the adapting population to successively higher peaks in the high-peaked region of the genotype space. If landscapes are very rugged and the high peaks do not cluster into smallish regions, recombination turns out to be a useless search strategy.

But most organisms are sexual. If organisms are sexual because recombination is a good search strategy, but recombination is only useful as a search strategy on certain classes of fitness landscapes, where did those fitness landscapes come from? No one knows.

The strange thing about evolution is that everyone thinks he understands it.

Somehow, evolution has brought forth the kind of smooth landscapes upon which recombination itself is a successful search strategy.

More generally, two young scientists, then at the Santa Fe Institute, proved a rather unsettling theorem. Bill Macready and David Wolpert called it the "no-free-lunch theorem." They asked an innocent question. Are there some search procedures that are "good" search procedures, no matter what the problem is? To formalize this, Bill and David considered a mathematical convenience—a set of all possible fitness landscapes. To be simple and concrete, consider a large three-dimensional room. Divide the room into very small cubic volumes, perhaps a millimeter on a side. Let the number of these small volumes in the room be large, say a trillion. Now consider all possible ways of assigning integers between one and a trillion to these small volumes. Any such assignment can be thought of as a fitness landscape, with the integer representing the fitness of that position in the room.

Next, formalize a search procedure as a process that somehow samples M distinct volumes among the trillion in the room. A search procedure specifies how to take the M samples. An example is a random search, choosing the M boxes at random. A second procedure starts at a box and samples its neighbors, climbing uphill via neighboring boxes toward higher integers. Still another procedure picks a box, samples neighbors, picks those with lower integers, and then continues.

The no-free-lunch theorem says that, averaged over all possible fitness landscape, no search procedure outperforms any other search procedure. What? Averaged over all possible fitness landscapes, you would do as well trying to find a large integer by searching randomly from an initial box for your M samples as you would climbing sensibly uphill from your initial box.

The theorem is correct. In the absence of any knowledge, or constraint, on the fitness landscape, on average, any search procedure is as good as any other.

But life uses mutation, recombination, and selection. These search procedures seem to be working quite well. Your typical bat or butterfly has managed to get itself evolved and seems a rather impressive entity. The no-free-lunch theorem brings into high relief the puzzle. If mutation, recombination, and selection only work well on certain kinds of fitness landscapes yet most organisms are sexual and hence use recombination, and all organisms use mutation as a search mechanism, where did these well-wrought fitness landscapes come from, such that evolution manages to produce the fancy stuff around us?

Here, I think, is how. Think of an organism's niche as a way of making a living. Call a way of making a living a "natural game." Then, of course, natural games evolve with the organisms making those livings during the past four billion years. What, then, are the "winning games"? Naturally, the winning games are the games the winning organisms play. One can almost see Darwin nod. But what games are those? What games are the games the winners play?

Ways of making a living, natural games, that are well searched out and well mastered by the evolutionary search strategies of organisms, namely, mutation and recombination, will be precisely the niches, or ways of making a living, that a diversifying and speciating population of organisms will manage to master. The ways of making a living presenting fitness landscapes that can be well searched by the procedures that organisms have in hand will be the very ways of making a living that readily come into existence. If there were a way of making a living that could not be well explored and exploited by organisms as they speciate, that way of making a living would not become populated. Good jobs, like successful jobholders, prosper.

So organisms, niches, and search procedures jointly and self-consistently coconstruct one another! We make the world in which we make a living such that we can, and have, more or less mastered that evolving world as we make it. The same is true, I will argue, for an econosphere. A web of eco-nomic activities, firms, tasks, jobs, workers, skills, and learning self-consistently came into existence in the last 40,000 years of human evolution.

The strange thing about the theory of evolution is that everyone thinks he understands it. But we do not. A biosphere, or an econosphere, self-consistently coconstructs itself according to principles we do not yet fathom.

Laws for a Biosphere

But there must be principles. Think of the Magna Carta, that cultural enterprise founded on a green meadow in England when John I was confronted by his nobles. British common law has evolved by precedent and determinations to a tangled web of more-or-less wisdom. When a judge makes a new determination, sets a new precedent, ripples of new interpretation pass to near and occasionally far reaches of the law. Were it the case that every new precedent altered the interpretation of all old judgments, the common law could not have coevolved into its rich tapestry. Conversely, if new precedents never sent out ripples, the common law could hardly evolve at all.

There must be principles of coevolutionary assembly for biospheres, economic systems, legal systems. Coevolutionary assembly must involve coevolving organizations flexible enough to change but firm enough to resist change. Edmund Burke was basically right. Might there be something deep here? Some hint of a law of coevolutionary assembly?

Perhaps. I begin with the simple example offered by Per Bak and his colleagues some years ago—Bak's "sand pile" and "self-organized criticality." The experiment requires a table and some sand. Drop the sand slowly on the table. The sand gradually piles up, fills the tabletop, and piles to the rest angle of sand, and then sand avalanches begin to fall to the floor.

Keep adding sand slowly to the sand pile and plot the size distribution of sand avalanches. You will obtain many small avalanches and progressively fewer large avalanches. In fact, you will achieve a characteristic size distribution called a "power law." Power law distributions are easily seen if one plots the logarithm of the number of avalanches at a given size on the y-axis, and the logarithm of the size of the avalanche on the x-axis. In the sand pile case, a straight line sloping downward to the right is obtained. The slope is the power law relation between the size and number of avalanches.

Bak and his friends called their sand pile "self-organized critical." Here, "critical" means that avalanches occur on all length scales, "self-organized" means that the system tunes itself to this critical state.

Many of us have now explored the application of Bak's ideas in models of coevolution. With caveats that other explanations may account for the data, the general result is that something may occur that is like a theory of coevolutionary assembly that yields a self-organized critical biosphere with a power law distribution of small and large avalanches of extinction and speciation events. The best data now suggest that precisely such a power law distribution of extinction and speciation events has occurred over the past 650 million years of the Phanerozoic. In addition, the same body of theory predicts that most species go extinct soon after their formation, while some live a long time. The predicted species lifetime distribution is a power law. So too are the data.

Similar phenomena may occur in an econosphere. Small and large avalanches of extinction and speciation events occur in our technologies. A colleague, Brian Arthur, is fond of pointing out that when the car came in, the horse, buggy, buggy whip, saddlery, smithy, and Pony Express went out of business. The car paved the way for an oil and gas industry, paved roads, motels, fast-food restaurants, and suburbia. The Austrian economist Joseph Schumpeter wrote about this kind of turbulence in capitalist economies. These Schumpeterian gales of creative destruction appear to occur in small and large avalanches. Perhaps the avalanches arise in power laws. And, like species, most firms die young; some make it to old age—Storre, in Sweden, is over nine hundred years old. The distribution of firm lifetimes is again a power law.

Here are hints—common law, ecosystems, economic systems—that general principles govern the coevolutionary coconstruction of lives and livings, organisms and natural games, firms and economic opportunities. Perhaps such a law governs any biosphere anywhere in the cosmos.

I have suggested other candidate laws for any biosphere in *Investigations*. As autonomous agents coconstruct a biosphere, each must manage to categorize and act upon its world in its own behalf. What principles might govern that categorization and action, one might begin to wonder. I suspect that autonomous agents coevolve such that each makes the maximum diversity of reliable discriminations on which it can act reliably as it swims, scrambles, pokes, twists, and pounces. This simple view leads to a working hypothesis: Communities of agents will coevolve to an "edge of chaos" between overrigid and overfluid behavior. The working hypothesis is richly testable today using, for example, microbial communities.

Moreover, autonomous agents forever push their way into novelty—molecular, morphological, behavioral, organizational. I will formalize this push into novelty as the mathematical concept of an "adjacent possible," persistently explored in a universe that can never, in the vastly many lifetimes of the universe, have made all possible protein sequences even once, bacterial species even once, or legal systems even once. Our universe is vastly nonrepeating; or, as the physicists say, the universe is vastly nonergodic. Perhaps there are laws that govern this nonergodic flow. I will suggest that a biosphere gates its way into the adjacent possible at just that rate at which its inhabitants can just manage to make a living, just poised so that selection sifts out useless variations slightly faster than those variations arise. We ourselves, in our biosphere, econosphere, and technosphere, gate our rate of discovery. There may be hints here too of a general law for any biosphere, a hoped-for new law for self-constructing systems of autonomous agents. Biospheres, on average, may enter their adjacent possible as rapidly as they can sustain; so too may econospheres. Then the hoped-for fourth law of thermodynamics for such self-constructing systems will be that they tend to maximize their dimensionality, the number of types of events that can happen next.

And astonishingly, we need stories. If, as I will suggest, we cannot prestate the configuration space, variables, laws, and initial and boundary conditions of a biosphere, if we cannot foretell a biosphere, we can, nevertheless, tell the stories as it unfolds. Biospheres demand their Shakespeares as well as their Newtons. We will have to rethink what science is itself. And C. P. Snow's "two cultures," the humanities and science, may find an unexpected, inevitable union.

FIVE

COMPLEXITY AND THE ARROW OF TIME

Paul Davies

The Dying Universe

In 1854, in one of the bleakest pronouncements in the history of science, the German physicist Hermann von Helmholtz claimed that the universe must be dying. He based his prediction on the second law of thermodynamics, according to which there is a natural tendency for order to give way to chaos. It is not hard to find examples in the world around us: people grow old, snowmen melt, houses fall down, cars rust, and stars burn out. Although islands of order may appear in restricted regions (e.g., the birth of a baby, crystals emerging from a solute) the disorder of the environment will always rise by an amount sufficient to compensate. This one-way slide into disorder is measured by a quantity called entropy. A state of maximum disorder corresponds to thermodynamic equilibrium, from which no change or escape is possible (except in the sense of rare statistical fluctuations). Helmholtz reasoned that the entropy of the universe as a whole remorselessly rises, presaging an end state in the far future characterized by universal equilibrium, following which nothing of interest will happen. This state was soon dubbed "the heat death of the universe."

Almost from the outset, the prediction of the cosmic heat death after an extended period of slow decay and degeneration was subjected to theological interpretation. The most famous commentary was given by the philosopher Bertrand Russell in his book *Why I Am Not a Christian*, in the following terms:

All the labors of the ages, all the devotion, all the inspiration, all the noonday brightness of human genius are destined to extinction in the vast death of the solar system, and the whole temple of man's achievement must inevitably be buried beneath the debris of a universe in ruins. All these things, if not quite beyond dispute, are yet so nearly certain that no philosophy which rejects them can hope to stand. Only within the scaffolding of these truths, only on the firm foundation of unyielding despair, can the soul's habitation henceforth be safely built. (Russell 1957, 107)

The association of the second law of thermodynamics with atheism and cosmic pointlessness has been an enduring theme. Consider, for example, this assessment by the British chemist Peter Atkins:

We have looked through the window on to the world provided by the Second Law, and have seen the naked purposelessness of nature. The deep structure of change is decay; the spring of change in all its forms is the corruption of the quality of energy as it spreads chaotically, irreversibly and purposelessly in time. All change, and time's arrow, point in the direction of corruption. The experience of time is the gearing of the electrochemical processes in our brains to this purposeless drift into chaos as we sink into equilibrium and the grave. (Atkins 1986, 98)

As Atkins points out, the rise of entropy imprints upon the universe an arrow of time, which manifests itself in many physical processes, the most conspicuous of which is the flow of heat from hot to cold; we do not encounter cold bodies getting colder and spontaneously giving up their heat to warm environments. The irreversible flow of heat and light from stars into the cold depths of space provides a cosmic manifestation of this simple "hot to cold" principle. On the face of it, this process will continue until the stars burn out and the universe reaches a uniform temperature. Our own existence depends crucially on the state of thermodynamic disequilibrium occasioned by this irreversible heat flow, since much life on Earth is sustained by the temperature gradient produced by sunshine. Microbes that live under the ground or on the sea bed utilize thermal and chemical gradients from the Earth's crust. These too are destined to diminish over time as thermal and chemical gradients equilibrate. Other sources of energy might provide a basis for life, but according to the second law, the supply of free energy continually diminishes until, eventually, it is all exhausted. Thus the death of the universe implies the death of all life, sentient or otherwise. It is probably this gloomy prognosis that led Steven Weinberg (1988) to pen the famous phrase "The more the universe seems comprehensible, the more it also seems pointless."

The fundamental basis for the second law is the inexorable logic of chance. To illustrate the principle involved, consider the simple example of a hot body in contact with a cold body. The heat energy of a material substance is due to the random agitation of its molecules. The molecules of the hot body move on average faster than those of the cold body. When the two bodies are in contact, the fast-moving molecules communicate some of their energy to the adjacent slow-moving molecules and speed them up. After a while, the higher energy of agitation of the hot body spreads across into the cold body, heating it up. In the end, this flow of heat brings the two bodies to a uniform temperature, in which the average energy of agitation is the same throughout. The flow of heat from hot to cold arises entirely because the chaotic molecular motions cause the energy to diffuse democratically among all the participating particles. The initial state, with the energy distributed in a lopsided way between the two bodies, is relatively more ordered than the final state, with the energy spread uniformly throughout the system. One way to see this is that a description of the initial state requires more information to specify it—namely, two numbers: the temperature of each body—whereas the final state can be described with only one number: the common final temperature. The loss of information occasioned by this transition may be quantified by the entropy of the system, which is roughly equal to the negative of the information content. Thus as information goes down, entropy, or disorder, goes up.

The transition of a collection of molecules from a low- to a high-entropy state is analogous to shuffling a deck of cards. Imagine that the cards are extracted from the package in suit and numerical order. After a period of random shuffling the cards will very probably be jumbled up. The transition from the initial ordered state to the final disordered one is due to the chaotic nature of the shuffling process. So the second law is really just a statistical effect of a rather trivial kind. It essentially says that a disordered state is much more probable than an ordered one—for the simple reason that there are numerically many more disordered states than ordered ones—so that when a system in an ordered state is randomly rearranged, it is very probably going to end up less ordered afterward. Thus blind chance lies at the basis of the second law of thermodynamics, just as it lies at the basis of Darwin's theory of evolution. Since chance, or contingency, as philosophers call it, is the opposite of law and order and hence purpose, it seems to offer powerful ammunition to atheists who wish to deny any overall cosmic purpose or design. If the universe is nothing but a physical system that began (for some mysterious reason) in a relatively ordered state and is inexorably shuffling itself into a chaotic one by the irresistible logic of probability theory, then it is hard to discern any overall plan or point.

Reaction to the Bleak Message
of the Second Law of Thermodynamics

Reaction to the theme of the dying universe set in already in the nineteenth century. Philosophers such as Henri Bergson (1964) and theologians like Teilhard de Chardin (Raven 1962) sought ways to evade or even refute the second law of thermodynamics. They cited evidence that the universe was in some sense getting better and better rather than worse and worse. In de Chardin's rather mystical vision, the cosmic destiny lay not with an inglorious heat death but with an enigmatic "omega point" of perfection. The progressive school of philosophy saw the universe as unfolding to ever greater richness and potential. Shortly afterward, the philosopher Alfred North Whitehead (1978) (curiously, the coauthor with Bertrand Russell of *Principia Mathematica*) founded the school of process theology on the notion that God and the universe are evolving together in a progressive rather than degenerative manner.

Much of this reaction to the second law had an element of wishful thinking. Many philosophers quite simply hoped and expected the law to be wrong. If the universe was apparently running down, like a heat engine running out of steam or a clock unwinding, then perhaps, they thought, nature has some process up its sleeve that can serve to wind it up again. Some sought this countervailing tendency in specific systems. For example, it was commonly supposed at the turn of the twentieth century that life somehow circumvents the strictures of thermodynamics and brings about increasing order. This was initially sought through the concept of vitalism—the existence of a life force that somehow bestowed order on the material contents of living systems. Vitalism eventually developed into a more scientific version, in what became known as organicism—the idea that complex organic wholes might have organizing properties that somehow override the trend into chaos predicted by thermodynamics (Sheldrake 1988). Others imagined that order could come out of chaos on a cosmic scale. This extended to periodic resurrections of the cyclic universe theory, according to which the entire cosmos eventually returns to some sort of pristine initial state after a long period of decay and degeneration. For example, in the 1960s it was suggested by the cosmologist Thomas Gold (1967) that one day the expanding universe may start to recontract and that during the contraction phase, the second law of thermodynamics would be reversed ("time will run backward"), returning the universe to a state of low entropy and high order. The speculation was based on a subtle misconception about the role of the expanding universe in the cosmic operation of the second law (discussed later). It turns out that the expansion of the universe crucially serves to provide the necessary thermodynamic disequilibrium that permits entropy to rise in the universe, but this does *not* mean a reversal of the expansion will cause a reversal of the entropic arrow. Quite the reverse: A rapidly contracting universe would drive entropy upward as effectively as a rapidly expanding one. In spite of this blind alley, the hypothesis that the directionality of physical processes might flip in a contracting universe was also proposed briefly by Stephen Hawking (Hawking, Laflamme, & Lyons 1993, 5342), who then abandoned the idea (Hawking 1994, 346), calling it his "greatest mistake." Yet the theory refuses to lie down. Recently, it was revived yet again by L. S. Schulman (1999).

The notion of a cyclic universe is, of course, an appealing one that is deeply rooted in many ancient cultures and persists today in Hinduism, Buddhism, and Aboriginal creation myths. The anthropologist Mircea Eliade (1954) termed it "the myth of the eternal return." In spite of detailed scrutiny, however, the second law of thermodynamics remains on solid scientific ground. So solid in fact that the astronomer Arthur Eddington felt moved to write, "if your theory is found to be against the second law of thermodynamics I can give you no hope; there is nothing for it but to collapse in deepest humiliation" (Eddington 1931, 447). Today, we know that there is nothing antithermodynamic about life. As for the cyclic universe theory, there is no observational evidence to support it (indeed, there is some rather strong evidence to refute it [Davies & Twamley 1993, 931]).

The True Nature of Cosmic Evolution

In this chapter, I wish to argue not that the second law is in any way suspect but that its significance for both theology and human destiny has been overstated. Some decades after Helmholtz's dying universe prediction, astronomers discovered that the universe is expanding. This changes the rules of the game somewhat. To give a simple example, there is good evidence that 300,000 years after the big bang that started the universe off, the cosmic matter was in a state close to thermodynamic equilibrium. This evidence comes from the detection of a background of thermal radiation that pervades the universe, thought to be the fading afterglow of the primeval heat. The spectrum of this radiation conforms exactly to that of equilibrium at a common temperature. Had the universe remained static at 300,000 years, it would in some respects have resembled the state of heat death described by Helmholtz. However, the expansion of the universe pulled the material out of equilibrium, allowing heat to flow and driving complex physical processes. The universe cooled as it expanded, but the radiation cooled more slowly than the matter, opening up a temperature gap and allowing heat to flow from one to the other. (The temperature of radiation when expanded varies in inverse proportion to the scale factor, whereas the temperature of nonrelativistic matter varies as the inverse square of the scale factor.) In many other ways too, thermodynamic disequilibrium emerged from equilibrium. This directionality is the "wrong way" from the point of view of a naïve application of the second law (which predicts a transition from disequilibrium to equilibrium) and shows that even as entropy rises, new sources of free energy are created.

I must stress that this "wrong way" tendency in no way conflicts with the letter of the second law. An analogy may be helpful to see why. Imagine a gas confined in a cylinder beneath a piston, as in a heat engine. The gas is in thermodynamic equilibrium at a uniform temperature. The entropy of the gas is a maximum. Now suppose that the gas is compressed by driving the piston forward; it will heat up as a consequence of Boyle's law. If the piston is now withdrawn again, restoring the gas to its original volume, the temperature will fall once more. In a reversible cycle of contraction and expansion, the final state of the gas will be the same as the initial state. What happens is that the piston must perform some work compressing the gas against its pressure, and this work appears as heat energy in the gas, raising its temperature. In the second part of the cycle, when the piston is withdrawn, the pressure of the gas pushes the piston out and returns exactly the same amount of energy as the piston injected. The temperature of the gas therefore falls to its starting value when the piston returns to its starting position.

However, for the cycle to be reversible, the motion of the piston must move very slowly relative to the average speed of the gas molecules. If the piston is moved suddenly, the gas will lag behind in its response, and this will cause a breakdown of reversibility. This is easy to understand. When the piston compresses the gas, if it moves fast there will be a tendency for the gas molecules to crowd up beneath the piston. As a result, the pressure of the gas beneath the piston will be slightly greater than the pressure within the body of the gas, so the piston will have to do rather more work to compress the gas than would have been the case had it moved more slowly. This will result in more energy being transferred from the advancing piston to the gas than would otherwise have been the case. Conversely, when the piston is suddenly withdrawn, the molecules have trouble keeping pace and lag back somewhat, thus reducing the density and pressure of the gas adjacent to the piston. The upshot is that the work done by the gas on the piston during the outstroke is somewhat less than the work done by the piston on the gas during the instroke. The overall effect is for a net transfer of energy from the piston to the gas, and the temperature, hence the entropy, of the gas rises with each cycle. Thus, although the gas was initially in a state of uniform temperature and maximum entropy, after the piston moves, the entropy nevertheless rises. The point being, of course, that to say the entropy of the gas is a maximum is to say that it has the highest value *consistent with the external constraints* of the system. But if those constraints change, by rapid motion of the piston, for example, then the entropy can go higher. During the movement phase, then, the gas will change from a state of equilibrium to a state of disequilibrium. This comes about not because the entropy of the gas falls—it never does— but because the maximum entropy of the gas increases, and, moreover, it increases faster than the actual entropy. The gas then races to "catch up" with the new constraints.

We can understand what is going on here by appreciating that the gas within a movable piston and cylinder is not an isolated system. To make the cycle run, there has to be an external energy source to drive the piston, and it is this source that supplies the energy to raise the temperature of the gas. If the total system—gas plus external energy source—is considered, then the system is clearly not in thermodynamic equilibrium to start with, and the rise in entropy of the gas is unproblematic. The entropy of the gas cannot go on rising forever. Eventually the energy source will run out, and the piston-and-cylinder device will stabilize in the final state of maximum entropy for the total system.

Where the confusion sets in is when the piston-and-cylinder expansion and contraction is replaced by the cosmological case of an expanding and (maybe, one day) contracting universe. Here the role of the piston-and-cylinder arrangement is played by the gravitational field. The external energy supply is provided by the gravitational energy of the universe. This has some odd features, because gravitational energy is actually negative. Think, for example, of the solar system. One would have to do work to pluck a planet from its orbit around the sun. The more material concentrates, the lower the gravitational energy becomes. Imagine a star that contracts under gravity; it will heat up and radiate more strongly, thereby losing heat energy and driving its gravitational energy more negative to pay for it. Thus the principle of a system seeking out its lowest energy state causes gravitating systems to grow more and more inhomogeneous with time. A smooth distribution of gas, for example, will grow clumpier with time under the influence of gravitational forces. Note that this is the opposite trend from a gas in which gravitation may be ignored. In that case the second law of thermodynamics predicts a transition toward uniformity. This is only one sense in which gravitation somehow goes "the wrong way."

It is tempting to think of the growth of clumpiness in gravitating systems as a special case of the second law of thermodynamics—that is, to regard the initial smooth state as a low-entropy (or ordered) state and the final clumpy state as a high-entropy (or disordered) one. It turns out that there are some serious theoretical obstacles to this simple characterization. One of these is that there seems to be no lower bound on the energy of the gravitational field. Matter can just go on shrinking to a singular state of infinite density, liberating an infinite amount of energy on the way. This fundamental instability in the nature of the gravitational field forbids any straightforward treatment of the thermodynamics of self-gravitating systems. In practice, an imploding ball of matter would form a black hole, masking the ultimate fate of the collapsing matter from view. So from the outside, there is a bound on the growth of clumpiness. We can think of the black hole as the equilibrium end state of a self-gravitating system. This interpretation was confirmed by Stephen Hawking, who proved that black holes are not strictly black but glow with thermal radiation (Hawking 1975, 199). The Hawking radiation has exactly the form corresponding to thermodynamic equilibrium at a characteristic temperature.

If we sidestep the theoretical difficulties of defining a rigorous notion of entropy for the gravitational field and take some sort of clumpiness as a measure of disorder, then it is clear that a smooth distribution of matter represents a low-entropy state as far as the gravitational field is concerned, whereas a clumpy state, perhaps including black holes, is a high-entropy state. Returning to the theme of the cosmic arrow of time, and the observed fact that the universe began in a remarkably smooth state, we may conclude that the matter was close to its maximum entropy state but the gravitational field was in a low-entropy state. The explanation for the arrow of time that describes the second law of thermodynamics lies therefore with an explanation for how the universe attained the smooth state it had at the big bang. Roger Penrose has attempted to quantify the degree of surprise associated with this smooth initial state (Penrose 1979, 581). In the case of a normal gas, say, there is a basic relationship between the entropy of its state and the probability that the state would be selected from a random list of all possible states. The lower the entropy, the less probable would be the state. This link is exponential in nature, so that as soon as one departs from a state close to equilibrium (i.e., maximum entropy) then the probability plummets. If one ignores the theoretical obstacles and just goes ahead and applies the same exponential statistical relationship to the gravitational field, it is possible to assess the "degree of improbability" that the universe should be found initially with such a smooth gravitational state. To do this, Penrose compared the actual entropy of the universe with the value it would have had if the big bang had coughed out giant black holes rather than smooth gas. Using Hawking's formula for the entropy of a black hole, Penrose was able to derive a discrepancy of 10^{30} between the actual entropy and the maximum possible entropy of the observable universe. Once this huge number is exponentiated, it implies a truly colossal improbability for the universe to start out in the observed relatively smooth state. In other words, the initial state of the universe is staggeringly improbable.

What should we make of this result? Should it be seen as evidence of design? Unfortunately, the situation is complicated by the inflationary universe scenario, which postulates that the universe jumped in size by a huge factor during the first split second. This would have the effect of smoothing out initial clumpiness. But this simply puts back the chain of explanation one step, because at some stage one must assume that the universe is in a less-than-maximum entropy state and hence an exceedingly improbable state. The alternative—that the universe began in its maximum entropy state—is clearly absurd, because it would then already have suffered the heat death.

The Cosmological Origin of Time's Arrow

The most plausible physical explanation for the improbable initial state of the universe comes from quantum cosmology, as expounded by Hawking, J. B. Hartle, and M. Gell-Mann (Hawking 1988). In this program, quantum mechanics is applied to the universe as a whole. The resulting "wave function of the universe" then describes its evolution. Quantum cosmology is beset with technical mathematical and interpretational problems, not least of which is what to make of the infinite number of different branches of the wave function, which describes a superposition of possible universes. The favored resolution is the many-universes interpretation, according to which each branch of the wave function represents a really existing parallel reality, or alternative universe.

The many-universes theory neatly solves the problem of the origin of the arrow of time. The wave function as a whole can be completely time-symmetric, but individual branches of the wave function will represent universes with temporal directionality. This has been made explicit in the time-symmetric quantum cosmology of Hartle and Gell-Mann, according to which the wave function of the universe is symmetric in time and describes a set of recontracting universes that start out with a big bang and end up with a big crunch (Gell-Mann & Hartle 1994, 311). The wave function is the same at each temporal extremity (bang and crunch). However, this does not mean that time runs backward in the recontracting phase of each branch, a là Thomas Gold. To be sure, there are some branches of the wave function where entropy falls in the recontracting phase, but these are exceedingly rare among the total ensemble of universes. The overwhelming majority of branches correspond to universes that either start out with low entropy and end up with high entropy, or vice versa. Because of the overall time symmetry, there will be equal proportions of universes with each direction of asymmetry. However, any observers in these universes will by definition call the low-entropy end of their universe the big bang and the high-entropy end the big crunch. Without the temporal asymmetry implied, life and observers would be impossible, so there is an anthropic selection effect, with those branches of the universe that are thermodynamically bizarre (starting and ending in equilibrium) going unseen. Thus, the ensemble of all possible universes shows no favored temporal directionality, although many individual branches do, and within those branches observers regard the "initial" cosmic states as exceedingly improbable. Although the Hartle and Gell-Mann model offers a convincing first step to explaining the origin of the arrow of time, it is not without its problems (see Davies & Twamley 1993).

To return to the description of our own universe (or our particular branch of the cosmological wave function), it is clear that the state of the universe in its early stages was one in which the matter and radiation were close to thermodynamic equilibrium but the gravitational field was very far from equilibrium. The universe started, so to speak, with its gravitational clock wound up but the rest in an unwound state. As the universe expanded, so there was a transfer of energy from the gravitational field to the matter, similar to that in the piston-and-cylinder arrangement. In effect, gravity wound up the rest of the universe. The matter and radiation started out close to maximum entropy consistent with the constraints, but then the constraints changed (the universe expanded). Because the rate of expansion was very rapid relative to the physical processes concerned, a lag opened up between the maximum possible entropy and the actual entropy, both of which were rising. In this way, the universe was pulled away from thermodynamic equilibrium by the expansion. Note that the same effect would occur if the universe contracted again, just as the instroke of the piston serves to raise the entropy of the confined gas. So there is no thermodynamic basis for supposing that the arrow of time will reverse should the universe start to contract.

The history of the universe, then, is one of entropy rising but chasing a moving target, because the expanding universe is raising the maximum possible entropy at the same time. The size of the entropy gap varies sharply as a function of time. Consider the situation at one second after the big bang. (I ignore here the situation before the first second, which was complicated but crucial in determining some important factors, such as the asymmetry between matter and antimatter in the universe.) The universe consisted of a soup of subatomic particles, such as electrons, protons and neutrons, and radiation. Apart from gravitons and neutrinos, which decoupled from the soup due to the weakness of their interactions well before the first second, the rest of the cosmic stuff was more or less in equilibrium. However, all this changed dramatically during the first 1,000 seconds or so. As the temperature fell, it became energetically favorable for protons and neutrons to stick together to form the nuclei of the element helium. All the neutrons got gobbled up this way, and about 25% of the matter was turned into helium. However, protons outnumbered neutrons, and most of the remaining 75% of the nuclear matter was in the form of isolated protons—the nuclei of hydrogen. Hydrogen is the fuel of the stars. It drives the processes that generate most of the entropy in the universe today, mainly by converting slowly into helium. So the lag behind equilibrium conditions is this: The universe would really "prefer" to be made of helium (it is more stable), but most of it is trapped in the form of hydrogen. I say trapped, because, after a few minutes, the temperature of the universe fell below that required for nuclear reactions to proceed, and it had to wait until stars were formed before the conversion of hydrogen into helium could be resumed. Thus the expansion of the universe generated a huge entropy gap—a gap between the actual and maximum possible entropy—during the first few minutes, when the equilibrium form of matter changed (due to the changing constraints occasioned by the cosmological expansion and the concomitant fall in temperature) from a soup of unattached particles to that of composite nuclei like helium. It was this initial few minutes that effectively "wound up the universe," giving it the stock of free energy and establishing the crucial entropy gap needed to run all the physical processes, like star burning, that we see today—processes that sustain interesting activity, such as life. The effect of starlight emission is to slightly close the entropy gap, but all the while the expanding universe serves to widen it. However, the rate of increase of the maximum possible entropy at our epoch is modest compared to what it was the first few minutes after the big bang—partly because the rate of expansion is much less, but also because the crucial nuclear story was all over in minutes. (The gap-generating processes occasioned by the expansion of the universe today are all of a less significant nature.) I haven't done the calculation, but I suspect that today the starlight entropy generation is more rapid than the rise in the maximum entropy caused by the expansion, so that the gap is probably starting to close, although it has a long way to go yet, and it could start to open up again if the dominant processes in the universe eventually proceed sufficiently slowly that they once more lag behind the pace of expansion. (This may happen if, as present observational evidence suggests, the expansion rate of the universe accelerates.)

The Ultimate Fate of the Universe

Of course one wants to know which tendency will win out in the end. Will the expanding universe continue to provide free energy for life and other processes, or will it fail to keep pace with the dissipation of energy by entropy-generating processes such as starlight emission? This question has been subjected to much study following a trail-blazing article by Freeman Dyson (Dyson 1979b, 447). First I should point out that the ultimate fate of the universe differs dramatically according to whether or not it will continue to expand forever. If there is enough matter in the universe it will eventually reach a state of maximum expansion, after which it will start to contract at an accelerating rate, until it meets its demise in a "big crunch" some billions of years later. In a simple model, the big crunch represents the end of space, time, and matter. The universe will not have reached equilibrium before the big crunch, so there will be no heat death but rather death by sudden obliteration. Theological speculations about the end state of a collapsing universe have been given by Frank Tipler (1994).

In the case of an ever-expanding universe, the question of its final state is a subtle matter. The outcome depends both on the nature of the cosmological expansion and on assumptions made about the ultimate structure of matter and the fundamental forces that operate in the universe. If the expansion remains more or less uniform, then when the stars burn out and nuclear energy is exhausted, gravitational energy from black holes could provide vast resources to drive life and other activity for eons. Over immense periods of time, even black holes evaporate via the Hawking effect. In one scenario, all matter and all black holes eventually decay, leaving space filled with a dilute background of photons and neutrinos, diminishing in density all the time. Nevertheless, as Dyson has argued, even in a universe destined for something like the traditional heat death, it is possible for the integrated lifetime of sentient beings to be limitless. This is because, as energy sources grow scarce, such beings could hibernate for longer and longer durations to conserve fuel. It is a mathematical fact that, by carefully husbanding ever-dwindling supplies of fuel, the total "awake" duration for a community of sentient beings may still be infinite.

Another possibility is that the expansion of the universe will not be uniform. If large-scale distortions, such as shear, become established, then they can supply energy through gravitational effects. Mathematical models suggest this supply of energy may be limitless. Still another scenario is that, from the dying remnants of this universe, another "baby" universe may be created using exotic quantum vacuum effects. Thus, even though our own universe may be doomed, others might be produced spontaneously, or perhaps artificially, to replace them as abodes for life. In this manner, the collection of universes, dubbed the "multiverse," may be eternal and ever-replenishing, even though any given universe is subject to the heat death. I have summarized these many and varied scenarios in my book *The Last Three Minutes* (see Davies 1994 and the references cited therein).

A Law of Increasing Complexity?

Clearly the heat death scenario of the nineteenth century does not well match the predicted fate of the universe in modern cosmology. But leaving this aside, the question arises as to the relevance of the second law of thermodynamics to cosmic change anyway. While it is undeniably true that the entropy of the universe increases with time, it is not clear that entropy is the most significant indicator of cosmic change. Ask a cosmologist for a brief history of the universe and you will get an answer along the following lines. In the beginning, the universe was in a very simple state—perhaps a uniform soup of elementary particles at a common temperature or even just expanding empty space. The rich and complex state of the universe as observed today did not exist at the outset. Instead it emerged in a long and complicated sequence of self-organizing and self-complexifying processes, involving symmetry-breaking, gravitational clustering, and the differentiation of matter. All this complexity was purchased at an entropic price. As I have explained, the rapid expansion of the universe just after the big bang created a huge entropy gap, which has been funding the accumulating complexification ever since and will continue to do so for a long while yet. Thus the history of the universe is not so much one of entropic degeneration and decay as a story of the progressive enrichment of systems on all scales, from atoms to galaxies.

It is tempting to postulate a universal principle of increasing complexity to go alongside the degenerative principle of the second law of thermodynamics. These two principles would not be in conflict, since complexity is not the negative of entropy. The growth of complexity can occur alongside the rise of entropy. Indeed, the time-irreversibility introduced into macroscopic processes by the second law of thermodynamics actually provides the opportunity for a law of increasing complexity, since if complexity were subject to time-reversible rules, there would then be a problem about deriving a time-asymmetric law from them. (This is a generalized version of the dictum that death is the price that must be paid for life.) In a dissipative system, an asymmetric trend toward complexity is unproblematic, and examples of such a trend are well known from the study of far-from-equilibrium (dissipative) processes (for a review, see Prigogine & Stengers 1984).

Many scientists have flirted with the idea of a quasi-universal law of increasing complexity, which would provide another cosmic arrow of time. For example, Freeman Dyson has postulated a principle of maximum diversity, according to which the universe in some sense works to maximize its richness, making it the most interesting system possible (Dyson 1979a, 250). In the more restrictive domain of biosystems, Stuart Kauffman has discussed a sort of fourth law of thermodynamics. In *Investigations*, he writes:

> I suspect that biospheres maximize the average secular construction of the diversity of autonomous agents, and ways those agents can make a living to propagate further. In other words, biospheres persistently increase the diversity of what can happen next. In effect, biospheres may maximize the average sustained growth of their own dimensionality. (2000, 3–4)

I have summarized various earlier attempts at formulating a principle of progressive cosmological organization in my book *The Cosmic Blueprint* (Davies 1987). These efforts bear a superficial resemblance to the medieval theological concept of the best of all possible worlds. However, they are intended to be physical principles, not commentaries on the human condition. It may well be the case that the laws of nature are constructed in such a way as to optimize one or more physical quantities, while having negligible implications for the specific activities of autonomous agents. Thus we can imagine that the laws of nature operate in such a way as to facilitate the emergence of autonomous agents and bestow up them the maximum degree of freedom consistent with biological and physical order, while leaving open the specifics of much of the behavior of those agents. So human beings are free to behave in undesirable ways, and the fact that human society may involve much misery—and is clearly not "the best of all possible worlds"—is entirely consistent with a principle of maximum richness, diversity, or organizational potential.

Attractive though the advance of complexity (or organization, or richness, or some other measure of antidegeneration) may be, there are some serious problems with the idea, not least that of arriving at a satisfactory definition of precisely what quantity it is that is supposedly increasing with time. A further difficulty manifests itself in the realm of biology. The evolution of the Earth's biosphere seems to be a prime example of the growth of organized complexity. After all, the biosphere today is far more complex than it was three and a half billion years ago. Moreover, extant individual organisms are more complex than the earliest terrestrial microbes. However, as Stephen Jay Gould has stressed, we must be very careful how to interpret these facts (Gould 1996). If life started out simple and randomly explored the space of possibilities, it is no surprise that it has groped on average toward increased complexity. But this is not at all the same as displaying a systematic trend or drive toward greater complexity.

To illustrate the distinction, I have drawn two graphs (fig. 5.1) showing possible ways in which the complexity of life might change with time. In both cases the mean complexity increases, but in figure 5.1a

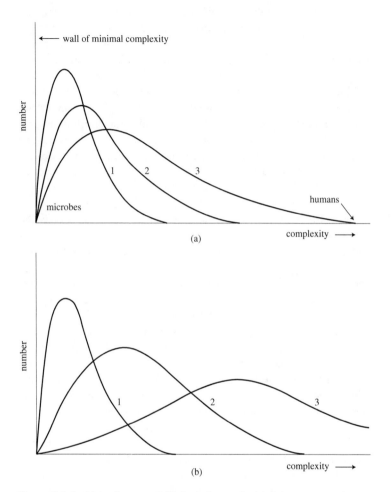

Figure 5.1 Ladder of progress? Biological complexity increases over time, but is there a systematic trend or just a random diffusion away from a "wall of simplicity"? The diffusion model, supported by Gould, is shown in (a). Curves 1, 2, and 3 represent successive epochs. Life remains dominated by simple microbes, but the tail of the distribution edges to the right. If there were a definite drive toward complexity, the curves would look more like in (b).

the increase is essentially just a diffusion effect, rather like a spreading wave packet in quantum mechanics. There is no progressive "striving" toward greater complexity. Clearly, the "wave packet" cannot diffuse to the left, because it is bounded by a state of minimal complexity (the least complexity needed for life to function). At any given time there will be a most complex organism, and it is no surprise if later most-complex organisms are more complex than earlier most-complex organisms, merely due to the spreading of the tail of the distribution. The fact that today, *homo sapiens* occupies the extreme end of the complexity distribution should not prejudice us into thinking that evolutionary history is somehow preoriented to the emergence of intelligence, or brain size, or whatever complex trait we deem significant. So the rise of complexity would be rather trivial in the case illustrated by figure 5.1a it does not illustrate any deep cosmic principle at work but merely the operation of elementary statistics. By contrast, figure 5.1b shows what a progressive trend towards greater complexity might look like. The "wave packet" still diffuses, but the median also moves steadily to the right, in the direction of greater complexity. If the latter description were correct, it would suggest that a principle of increasing complexity were at work in biological evolution.

What does the fossil record show? Gould (1996) argues forcefully that it supports figure 5.1a and confounds figure 5.1b. His is the standard neo-Darwinian position. It is the essence of Darwinism that nature has no foresight—it cannot look ahead and anticipate the adaptive value of specific complex features. By contrast, the concept of "progressive" trends has an unnervingly teleological, even Lamarckian, flavor to it, making it anathema to most biologists (see, for example, Nitecki 1988). On the experimental front, however, the fossil record is scrappy, and there have been challenges to Gould's position. For example, it has been pointed out that the rate of encephalization (brain to body mass ratio) is on an accelerating trend (Russell 1993, 95). Moreover, the simplest microbes known seem to be far more complex than the minimal possible, although this is open to the objection that we may not yet have identified the simplest autonomous organisms.

Many SETI researchers unwittingly support figure 5.1b in anticipating that intelligence and advanced technological communities will arise more or less automatically once life gets started. Not surprisingly, perhaps, Gould, in common with many biologists, is a critic of SETI, believing that intelligence is exceedingly unlikely to develop on another planet, as there is no evolutionary trend, he maintains, that canalizes complexity in the direction of large brains.

The clearest example of a trend toward complexity comes not from biology but astrophysics. A snapshot of the early universe was provided by the COBE satellite, which mapped tiny variations in the heat radiation left over from the big bang. This radiation is extremely smooth across the sky, indicating that the early universe was highly uniform, as I discussed earlier. The amplitude of the temperature ripples that COBE detected are about one part per hundred thousand only. The hot spots represent the first glimmerings of large-scale cosmological structure. Over time, the hotter, denser regions of the universe have grown at the expense of the rest. Gravity has caused matter to accumulate into galaxies and the galaxies into clusters. Within galaxies, matter has concentrated into stars. It seems intuitively obvious that the elaborate structure of the Milky Way galaxy, with its spiral arms and myriad stars, is far more complex than the diffuse swirling nebula that produced it.

However, quantifying this growth in complexity is hard. Over the last few decades there have been several attempts to try to develop a measure of gravitational complexity that will describe the irreversible trend in gravitating systems from simple smooth initial states to more complex, clumpy final states. As already explained, the end state resulting from this gravitational concentration of matter is the black hole. One idea is to try and describe the growth of clumpiness in terms of entropy, since it would then have an explanation in a suitably generalized second law of thermodynamics. Penrose has made some specific proposals along these lines (see Penrose 1979). Unfortunately, no satisfactory definition of gravitational entropy has emerged.

What sort of principle would a law of increasing complexity be? First, it is obvious that there can be no absolute and universal law of this sort, because it is easy to think of counterexamples. Suppose Earth were hit tomorrow by a large asteroid that wiped out all "higher" life forms? That event would reset the biological complexity of the planet to a seriously lower level. However, we might still be able to show that, following such a setback, and otherwise left alone, life rides an escalator of complexity growth. Another caveat concerns the ultimate fate of the universe. So long as there is an entropy gap, the universe can go on creating more and more complexity. However, in some scenarios of the far future, matter and black holes eventually decay and evaporate away, leaving a very dilute soup of weakly interacting subatomic particles gradually cooling toward absolute zero and expanding to a state of zero density. The complexity of such a state seems much lower than today (although the presently observable universe will have grown to an enormous volume by then, so the total complexity may still be rather large). Therefore, in the end, complexity may decline. However, this does not preclude a more restrictive law of increasing complexity—one subject to certain prerequisites.

If some sort of complexity law were indeed valid, what would be its nature? Complexity theorists are fond of saying that they seek quasi-universal principles that would do for complexity what the laws of thermodynamics do for entropy. But would such a law emerge as a *consequence* of known physics, or do the laws of complexity *complement* the laws of physics? Are they emergent laws? Robert Wright has argued that the history of evolution drives an increase in complexity through the logic of non-zero-sumness—that is, individual units, whether they are cells, animals, or communities, may cooperate in ways that produce a net benefit for all (Wright 2000). Evolution will exploit such win-win interactions and select for them. The question is then whether the non-zero-sum principle is something that augments the laws of nature or follows from them. If it augments them, whence comes this higher-level law? If it is a consequence of the laws of nature, then will such a principle follow from a random set of natural laws, or is there something special about the actual laws of the real universe that facilitates the emergence of non-zero-sum interactions?

Whatever the answers to these difficult questions, it is clear that any general principle of advancing complexity would reintroduce a teleological element into science, against the trend of the last three hundred years. This is not a new observation. A hint of teleology is discernible in the approach of many investigators who have suggested that complexity may grow with time.[1] For example, Kauffman writes of "we, the expected" (Kauffman 1995). But I shall finish with my favorite quotation of Freeman Dyson, who has said, poetically, that "in some sense the universe must have known we were coming" (Dyson 1979b, 250).

NOTE

1. It is, in fact, possible to give expression to apparent "anticipatory" qualities in nature without reverting to causative teleology in the original sense. I have discussed this topic in a recent article (see Davies 1998, 151).

REFERENCES

Atkins, Peter. 1986. "Time and Dispersal: The Second Law." In *The Nature of Time*, edited by Raymond Flood and Michael Lockwood. Oxford: Blackwell.

Bergson, Henri. 1964. *Creative Evolution*. Translated by A. Mitchell. London: Macmillan.

Davies, Paul. 1987. *The Cosmic Blueprint*. London: Heinemann.

Davies, Paul. 1994. *The Last Three Minutes*. New York: Basic Books.

Davies, Paul. 1998. "Teleology without Teleology: Purpose through Emergent Complexity." In *Evolutionary and Molecular Biology: Scientific Perspectives on Divine Action*, edited by R. Russell, W. R. Stoeger, and S. J. and F. J. Ayala. Vatican City: Vatican Observatory.

Davies, P. C. W., and J. Twamley. 1993. "Time Symmetric Cosmology and the Opacity of the Future Light Cone." *Classical and Quantum Gravity* 10: 931.

Dyson, Freeman. 1979a. *Disturbing the Universe*. New York: Harper and Row.

Dyson, Freemann. 1979b. "Time without End: Physics and Biology in an Open Universe." *Reviews of Modern Physics* 51: 447–60.

Eddington, A. S. 1931. "The End of the World from the Standpoint of Mathematical Physics." *Nature* 127: 447.

Eliade, Mircea. 1954. *The Myth of Eternal Return*. Translated by W. R. Trask. New York: Pantheon Books.

Gell-Mann, M., and J. B. Hartle. 1994. "Time Symmetry and Asymmetry in Quantum Mechanics and Quantum Cosmology." In *Physical Origins of Time Asymmetry*, edited by. J. J. Halliwell, J. Perez-Mercader, and W. H. Zurek. Cambridge: Cambridge University Press.

Gold, Thomas, ed. 1967. *The Nature of Time*. Ithaca: Cornell University Press.

Gould, Stephen Jay. 1996. *Life's Grandeur*. London: Jonathan Cape.

Hawking, S. W. 1975. "Particle Creation by black holes." *Communications in Mathematical Physics* 43.

Hawking, S. W. 1988. *A Brief History of Time*. New York: Bantam Books.

Hawking, S. W. 1994. "The No Boundary Condition and the Arrow of Time." In *Physical Origins of Time Assymetry*, edited by J. J. Halliwell, J. Perez-Mercader, and W. H. Zurek. Cambridge: Cambridge University Press.

Hawking, S. W., R. Laflamme, and G. W. Lyons. 1993. "The Origin of Time Asymmetry." *Physical Review* D47.

Kaufmann, Stuart. 1995. *At Home in the Universe*. Oxford: Oxford University Press.

Kaufmann, Stuart. 2000. *Investigations*. Oxford: Oxford University Press.

Nitecki, Matthew H. 1988. *Evolutionary Progress*. Chicago: University of Chicago Press.

Penrose, Roger. 1979. "Singularities and Time Asymmetry." In *General Relativity: An Einstein Centenary Survey*, edited by S. W. Hawkins and W. Israel. Cambridge: Cambridge University Press.

Prigogine, Ilya, and Isabelle Stengers. 1984. *Order out of Chaos*. London: Heinemann.

Raven, C. 1962. *Teilhard de Chardin: Scientist and Seer*. New York: Harper and Row.

Russell, Bertrand. 1957. *Why I Am Not a Christian*. New York: Allen and Unwin.

Russell, James Dale. 1993. "Exponential Evolution: Implications for Intelligent Extraterrestrial Life." *Advances in Space Research* 3: 95.

Schulman, L. S. 1999. "Opposite Thermodynamic Arrows of Time." *Physical Review Letters* 83.

Sheldrake, Rupert. 1988. *The Presence of the Past*. London: Collins.

Tipler, Frank. 1994. *The Physics of Immortality*. New York: Doubleday.

Weinberg, Steven. 1988. *The First Three Minutes*. Updated ed. New York: Harper and Row.

Whitehead, Alfred North. 1978. *Process and Reality: An Essay in Cosmology*. Corrected ed. Edited by. D. R. Griffin and D. W. Sherburne. New York: Free Press.

Wright, Robert. 2000. *Nonzero*. New York: Pantheon.

SIX

CAN EVOLUTIONARY ALGORITHMS GENERATE SPECIFIED COMPLEXITY?

William A. Dembski

Davies's Challenge

In *The Fifth Miracle* Paul Davies suggests that any laws capable of explaining the origin of life must be radically different from scientific laws known to date (Davies 1999). The problem, as he sees it, with currently known scientific laws, like the laws of chemistry and physics, is that they cannot explain the key feature of life that needs to be explained; that feature is *specified complexity*. Life is both complex and specified. The basic intuition here is straightforward. A short word like the definite article "the" is specified without being complex (it conforms to an independently given pattern but is simple). A long sequence of random letters is complex without being specified (it requires a complicated instruction-set to characterize but conforms to no independently given pattern). A Shakespearean sonnet is both complex and specified.

Now, as Davies rightly notes, *laws* (i.e., necessities of nature) can explain specification but not complexity. For instance, the formation of a salt crystal follows well-defined laws, produces an independently given repetitive pattern, and is therefore specified; but that pattern will also be simple, not complex. On the other hand, as Davies also rightly notes, *contingency* (i.e., chance or accidental processes of nature) can explain complexity but not specification. For instance, the exact time sequence of radioactive emissions from a piece of uranium will be contingent and complex but not specified. The problem is to explain something like the genetic code, which is both complex and specified. As Davies puts it: "Living organisms are mysterious not for their complexity *per se*, but for their tightly specified complexity" (Davies 1999, 112).

Specified complexity is a type of information. Indeed, both complexity and specification are well-defined information-theoretic concepts. Complexity here is used in the Shannon sense and denotes a measure of improbability. Take, for instance, a combination lock: The more possible combinations of the lock, the more complex the mechanism and correspondingly the more improbable that the mechanism can be opened by chance. Complexity and probability therefore vary inversely—the greater the complexity, the smaller the probability. Specification here refers to the patterning of complex arrangements where the pattern can be recovered independently of the actual arrangement. Specified complexity admits a rigorous mathematical formulation.

How then does the scientific community explain specified complexity? Usually via an evolutionary algorithm. By an evolutionary algorithm I mean any well-defined mathematical procedure that generates contingency via some chance process and then sifts it via some lawlike process. The Darwinian mechanism, simulated annealing, neural nets, and genetic algorithms all fall within this broad definition of evolutionary algorithms. Evolutionary algorithms constitute the mathematical underpinnings of Darwinian theory.

Now, the problem with invoking evolutionary algorithms to explain specified complexity at the origin of life is the absence of any identifiable evolutionary algorithm that might account for it. Once life has started and self-replication has begun, the Darwinian mechanism is usually invoked to explain the specified complexity of living things. But what is the relevant evolutionary algorithm that drives chemical evolution? No convincing answer has been given to date. To be sure, one can hope that an evolutionary algorithm that generates specified complexity at the origin of life exists and remains to be discovered. Manfred Eigen, for instance, writes (Eigen 1992, 12), "Our task is to find an algorithm, a natural law that leads to the origin of information," where by "information" he means specified complexity. But if some evolutionary algorithm can be found to account for the origin of life, it would not be a radically new law in Davies's sense. Rather, it would be a special case of a known process.

I submit that the problem of explaining specified complexity is even worse than Davies makes out in *The Fifth Miracle*. Not only have we yet to explain specified complexity at the origin of life, but evolutionary algorithms fail to explain it in the subsequent history of life as well. Given the popular enthusiasm for evolutionary algorithms, such a claim may seem misconceived. But consider a well-known example by Richard Dawkins in which he purports to show how an evolutionary algorithm can generate specified complexity (Dawkins 1986, 47–48).

He starts with a target sequence taken from Shakespeare's *Hamlet*, namely, METHINKS IT IS LIKE A WEASEL. If we tried to attain this sequence by pure chance (for example, by randomly shaking out Scrabble pieces), the probability of getting it on the first try would be around 1 in 10^{40}, and correspondingly it would take on average about 10^{40} tries to stand a better-than-even chance of getting it. Thus, if we depended on pure chance to attain this target sequence, we would in all likelihood be unsuccessful. As a problem for pure chance, attaining Dawkins's target sequence is an exercise in generating specified complexity, and it becomes clear that pure chance simply is not up to the task.

But consider next Dawkins's reframing of the problem. In place of pure chance, he considers the following evolutionary algorithm. (1) Start with a randomly selected sequence of 28 capital Roman letters and spaces (that's the length of METHINKS IT IS LIKE A WEASEL); (2) randomly alter all the letters and spaces in the current sequence that do not agree with the target sequence; and (3) whenever an alteration happens to match a corresponding letter in the target sequence, leave it and randomly alter only those remaining letters that still differ from the target sequence. In very short order this algorithm converges to Dawkins's target sequence. In *The Blind Watchmaker*, Dawkins recounts a computer simulation of this algorithm that converges in 43 steps. In place of 10^{40} tries on average for pure chance to generate the target sequence, it now takes on average only 40 tries to generate it via an evolutionary algorithm.

Dawkins and fellow Darwinists use this example to illustrate the power of evolutionary algorithms to generate specified complexity. Closer investigation, however, reveals that Dawkins's algorithm has not so much generated specified complexity as shuffled it around. As I will show, invariably when evolutionary algorithms appear to generate specified complexity, what they actually do is smuggle in preexisting specified complexity. Indeed, evolutionary algorithms are inherently incapable of generating specified complexity. The rest of this chapter is devoted to justifying this claim.

Statement of the Problem

Generating specified complexity via an evolutionary algorithm is typically understood as an optimization problem. Optimization is one of the main things mathematicians do for a living. What's the shortest route connecting 10 cities? What's the fastest algorithm to sort through a database? What's the most efficient way to pack objects of one geometrical shape inside another (e.g., bottles in boxes)? Optimization quantifies superlatives like "shortest," "fastest," and "most efficient" and attempts to identify states of affairs to which those superlatives apply.

Generating specified complexity via an evolutionary algorithm can be understood as the following optimization problem. We are given a reference class of possible solutions known as the *phase space*. Possible solutions within the phase space are referred to as *points*. A univalent measure of optimality known as the *fitness function* is then defined on the phase space. The fitness function is a nonnegative real-valued function that is optimized by being maximized. The task of an evolutionary algorithm is to locate where the fitness function attains at least a certain level of fitness. The set of possible solutions where the fitness function attains at least that level of fitness will be called the *target*.

Think of it this way. Imagine that the phase space is a vast plane and the fitness function is a vast hollowed-out mountain range over the plane (complete with low-lying foothills and incredibly high peaks). The task of an evolutionary algorithm is by moving around on the plane to get to some point under the mountain range where it attains at least a certain height (say 10,000 feet). The collection of all places on the plane where the mountain range attains at least that height is the target. Thus the job of the evolutionary algorithm is to find its way into the target by navigating the phase space.

Now the phase space (which we are picturing as a giant plane) invariably comes with some additional topological structure, typically given by a metric or distance function. A metric assigns to any pair of points the distance separating those points. The topological structure induced by a metric tells us how points in the phase space are related geometrically to nearby points. Though typically huge, phase spaces tend to be finite (strictly finite for problems represented on computer and topologically finite, or what topologists call "compact," in general). Moreover, such spaces typically come with a uniform probability that is adapted to the topology of the phase space. A uniform probability assigns identical probabilities to geometrically congruent pieces of phase space.

If you think of the phase space as a giant plane, this means that if you get out your tape measure and measure off, say, a 3-by-5-foot area in one part of the phase space, the uniform probability will assign it the same probability as a 3-by-5-foot area in another portion of the phase space. All the spaces to which evolutionary algorithms have until now been applied do indeed satisfy these two conditions of having a finite topological structure (i.e., they are compact) and possessing a uniform probability. Moreover, this uniform probability is what typically gets used to estimate the complexity or improbability of the target.

For instance, in Dawkins's METHINKS IT IS LIKE A WEASEL example, the phase space consists of all sequences of upper-case Roman letters and spaces 28 characters in length. In general, given a phase space whose target is where a fitness function attains a certain height, the uniform probability of randomly choosing a point from the phase space and landing in the target will be extremely small. In Dawkins's example, the target is the character string METHINKS IT IS LIKE A WEASEL and the improbability is 1 in 10^{40}. For nontoy examples the improbability is typically much more extreme than that. Note that if the probability of the target were not small, a random or blind search of the phase space would suffice to locate the target, and there would be no need to construct an evolutionary algorithm.

We therefore suppose that the target has minuscule probability. Moreover, we suppose that the target, in virtue of its explicit identification, is specified (certainly this is the case in Dawkins's example, for which the target coincides with a line from Shakespeare's *Hamlet*). Thus it would seem that for an evolutionary algorithm to locate the target would be to generate specified complexity.

But let's look deeper. An evolutionary algorithm is a stochastic process, that is, an indexed set of random values with precisely given probabilistic dependencies. The values an evolutionary algorithm assumes occur in phase space. An evolutionary algorithm moves around phase space some finite number of times and stops once it reaches the target. The indexing set for an evolutionary algorithm is therefore the natural numbers 1, 2, 3, and so on. Since we need the evolutionary algorithm to locate the target in a manageable number of steps, we fix a natural number m and consider only those values of algorithm indexed by 1 through m (we refer to m as the "sample size").

An evolutionary algorithm needs to improve on blind search. In most cases of interest the target is so small that the probability of finding it in m steps via a blind search is minuscule. On the other hand, if the evolutionary algorithm is doing its job, its probability of locating the target in *m* steps will be quite large. And since throughout this discussion complexity and improbability are equivalent notions, the target, though complex and specified with respect to the uniform probability on the phase space, remains specified but no longer complex with respect to the evolutionary algorithm.

But doesn't this mean that the evolutionary algorithm has generated specified complexity after all? No. At issue is the *generation* of specified complexity, not its *reshuffling*. To appreciate the difference, one must be clear about a condition the evolutionary algorithm must satisfy if it is to count as a legitimate correlative of the Darwinian mutation-selection mechanism. It is not, for instance, legitimate for the evolutionary algorithm to survey the fitness landscape induced by the fitness function, see where in the phase space it attains a global maximum, and then head in that direction. That would be teleology. No, the evolutionary algorithm must be able to navigate its way to the target either by randomly choosing points from the phase space or by using those as starting points and then selecting other points based solely on the topology of the phase space and without recourse to the fitness function, except to evaluate it at points already traversed by the algorithm. In other words, the algorithm must move around the phase space only on the basis of its topology and the elevation of the fitness function at points in the phase space already traversed.

Certainly this means that the evolutionary algorithm has to be highly constrained in its use of the fitness function. But there's more. It means that its success in hitting the target depends crucially on the structure of the fitness function. If, for instance, the fitness function is totally flat and close to zero whenever it is outside the target, then it fails to discriminate between points outside the target and so cannot be any help guiding an evolutionary algorithm into the target. For such a fitness function, the probability of the evolutionary algorithm landing in the target is no better than the probability of a blind search landing in the target (an eventuality we've dismissed out of hand—the target simply has too small a probability for blind search to stand any hope of success). I will now turn in detail to the fitness function and crucial role its structure plays in guiding an evolutionary algorithm into the target.

The Structure of the Fitness Function

To understand how a fitness function guides an evolutionary algorithm into a target, consider again Dawkins's METHINKS IT IS LIKE A

WEASÈL example. Recall that the phase space in this example comprised all sequences of capital Roman letters and spaces 28 characters in length and that the target was the sentence METHINKS IT IS LIKE A WEASEL. In this example the essential feature of the fitness function was that it assigned higher fitness to sequences having more characters in common with the target sequence. There are many ways to represent such a fitness function mathematically, but perhaps the simplest is simply to count the number of characters identical with the target sequence. Such a fitness function ranges between 0 and 28, assigning 0 to sequences with no coinciding characters and 28 solely to the target sequence.

Dawkins's evolutionary algorithm started with a random sequence of 28 characters and then at each stage randomly modified all the characters that did not coincide with the corresponding character in the target sequence. As we've seen, this algorithm converged on the target sequence with probability 1 and on average yielded the target sequence after about 40 iterations. Now, when the algorithm reaches the target, has it in fact generated specified complexity? To be sure, not in the sense of generating a target sequence that is inherently improbable or complex for the algorithm (the evolutionary algorithm converges to the target sequence with probability 1). Nonetheless, with respect to the original uniform probability on the phase space, which assigned to each sequence a probability of around 1 in 10^{40}, the evolutionary algorithm appears to have done just that, to wit, generate a highly improbable specified event, or what we are calling specified complexity.

Even so, closer inspection reveals that specified complexity, far from being generated, has merely been smuggled in. Indeed, it is utterly misleading to say that Dawkins's algorithm has *generated* specified complexity. What the algorithm has done is take advantage of the specified complexity inherent in the fitness function and utilize it in searching for and then locating the target sequence. Any fitness function that assigns higher fitness to sequences the more characters they have in common with the target sequence is hardly going to be arbitrary. Indeed, it is going to be *highly specific* and *carefully adapted* to the target. Indeed, its definition is going to require more complex specified information than the original target.

This is easily seen in the example. Given the sequence METHINKS IT IS LIKE A WEASEL, we adapted the fitness function to this sequence so that the function assigns the number of places where an arbitrary sequence agrees with it. But note that in the construction of this fitness function, there is nothing special about the sequence

METHINKS IT IS LIKE A WEASEL. Any other character sequence of 28 letters and spaces would have served equally well. Given any target sequence whatsoever, we can define a fitness function that assigns the number of places where an arbitrary sequence agrees with it. Moreover, given this fitness function, our evolutionary algorithm will just as surely converge to the new target as previously it converged to METHINKS IT IS LIKE A WEASEL.

It follows that every sequence corresponds to a fitness function specifically adapted for conducting the evolutionary algorithm into it. Every sequence is therefore potentially a target for the algorithm, and the only thing distinguishing targets is the choice of fitness function. But this means that the problem of finding a given target sequence has been *displaced* to the new problem of finding a corresponding fitness function capable of locating the target. The original problem was finding a certain target sequence within phase space. The new problem is finding a certain fitness function within the entire collection of fitness functions—and one that is specifically adapted for locating the original target.

The collection of all fitness functions has therefore become a new phase space in which we must locate a new target (the new target being a fitness function capable of locating the original target in the original phase space). But this new phase is far less tractable than the original phase space. The original phase space comprised all sequences of capital Roman letters and spaces 28 characters in length. This space contained about 10^{40} elements. The new phase space, even if we limit it, as I have here, to integer-valued functions with values between 0 and 28, will have at least $10^{10^{40}}$ elements (i.e., one followed by 10^{40} zeros). If the original phase space was big, the new one is a monster.

To say that the evolutionary algorithm has generated specified complexity within the original phase space is therefore really to say that it has borrowed specified complexity from a higher-order phase space, namely, the phase space of fitness functions. And since this phase space is always much bigger and much less tractable than the original phase space, it follows that the evolutionary algorithm has in fact not generated specified complexity at all but merely shifted it around.

We have here a particularly vicious regress. For the evolutionary algorithm to generate specified complexity within the original phase space presupposes that specified complexity was first generated within the higher-order phase space of fitness functions. But how was this prior specified complexity generated? Clearly, it would be self-defeating to claim that some higher-order evolutionary algorithm on the higher-order phase space of fitness functions generated specified complexity;

for then we face the even more difficult problem of generating specified complexity from a still higher-order phase space (i.e., fitness functions over fitness functions over the original phase space).

This regress, in which evolutionary algorithms shift the problem of generating specified complexity from an original phase space to a higher-order phase space, holds not just for Dawkins's METHINKS IT IS LIKE A WEASEL example but in general. Complexity theorists are aware of this regress and characterize it by what are called no-free-lunch theorems. The upshot of these theorems is that averaged over all possible fitness functions, no search procedure outperforms any other. It follows that any success an evolutionary algorithm has in outputting specified complexity must ultimately be referred to the fitness function that the evolutionary algorithm employs in conducting its search.

The No-Free-Lunch Theorems

The no-free-lunch theorems dash any hope of generating specified complexity via evolutionary algorithms. Before turning to them, however, it is necessary to consider blind search more closely. Joseph Culberson begins his discussion of the no-free-lunch (NFL) theorems with the following vignette.

> In the movie *UHF*, there is a marvelous scene that every computing scientist should consider. As the camera slowly pans across a small park setting, we hear a voice repeatedly asking "Is this it?" followed each time by the response "No!" As the camera continues to pan, it picks up two men on a park bench, one of them blind and holding a Rubik's cube. He gives it a twist, holds it up to his friend and the query-response sequence is repeated. This is blind search. (Culberson 1998, 109)

Blind search can be characterized as follows. Imagine two interlocutors, Alice and Bob. Alice has access to a reference class of possible solutions to a problem—what I'm calling the phase space. Bob has access not only to the phase space but also to the set of actual solutions, which I'm calling the target. For any possible solution in the phase space, Bob is able to tell Alice whether it falls in the target (that is, whether it is in fact a solution). Alice now successively selects *m* possible solutions from the phase space, at each step querying Bob (note that Alice limits herself to a finite search with at most *m* steps—she does not have infinite resources to continue the search indefinitely). Bob then truthfully answers whether each proposed solution is in fact in the target. The search is successful if one of the proposed solutions lands in the target. Computer scientists call this blind search.

The crucial question now is this: Given that the target is so improbable that blind search is highly unlikely to succeed, what additional information might help Alice to make her search succeed? To answer this question, return to the exchange between Alice and Bob. Alice and Bob are playing a game of "m questions" in which Bob divulges too little information for Alice to have any hope of winning the game. Alice therefore needs some additional information from Bob. But what? Bob could just inform Alice of the exact location of the target and be done with it. But that would be too easy. If Alice is playing the role of scientist and Bob the role of nature, then Bob needs to make Alice drudge and sweat to locate the target—nature, after all, does not divulge her secrets easily. Alice and Bob are operating here with competing constraints. Bob wants to give Alice the minimum information she needs to locate the target. Alice, on the other hand, wants to make maximal use of whatever information Bob gives her to ensure that her m questions are as effective as possible in locating the target.

Suppose, therefore, that Bob identifies some additional information and makes it available to Alice. This information is supposed to help Alice locate the target. There is therefore a new protocol for the exchange between Alice and Bob. Before, Bob would only tell Alice whether a candidate solution belonged to the target. Now, for any candidate solution that Alice proposes, Bob will tell her what this additional information has to say about it. We therefore have a new game of "m questions" in which the answer to each question is not *whether* some proposed solution belongs to the target but rather *what* the additional information has to say specifically about each candidate solution. It follows that all the action in this new game of "m questions" centers on the additional information. Is it enough to render Alice's m-step search for the target successful? And if so, what characteristics must this additional information possess?

The brilliant insight behind the no-free-lunch theorems is that it doesn't matter what characteristics the additional information possesses. Instead, what matters is the reference class of possibilities to which this information belongs and from which it is drawn. Precisely because it is information, it must by definition belong to a reference class of possibilities. Information always presupposes an informational context of multiple live possibilities. Robert Stalnaker puts it this way (Stalnaker 1984, 85): "Content requires contingency. To learn something, to acquire information, is to rule out possibilities. To understand the information conveyed in a communication is to know what possibilities would be excluded by its truth." Fred Dretske elaborates (Dretske 1981, 4): "Information theory identifies the amount of information associated with, or generated by, the occurrence of an event (or the realization of a state of affairs) with the reduction in uncertainty, the elimination of possibilities, represented by that event or state of affairs."

I shall, therefore, refer to the reference class of possibilities from which the additional information is drawn as the *informational context*. The informational context constitutes the totality of informational resources that might assist Alice in locating the target. We suppose that Bob has full access to the informational context, selects some item of information from it, and makes it available to Alice to help her locate the target. The reason we speak of no-free-lunch theorem*s* (plural) is to distinguish the different types of informational contexts from which Bob might select information to assist Alice.

What can the informational context look like? I've already shown that the informational context is the class of fitness functions on the phase space. Since the phase space is a topological space, the informational context could as well be the continuous fitness functions on the phase space. If the phase space is a differentiable manifold, the informational context could be the differentiable fitness functions on the phase space. The informational context could even be a class of temporally indexed fitness functions that identify not merely fitness but fitness at some time t (such temporally indexed fitness functions have yet to find widespread use but seem much more appropriate for modeling fitness, which in any realistic environment is not likely to be static). Alternatively, the informational context need not involve any fitness functions whatsoever. The informational context could be a class of dynamical systems, describing the flow of particles through phase space. The possibilities for such informational contexts are limitless, and each such informational context has its own no-free-lunch theorem.

A generic no-free-lunch theorem now looks as follows: It sets up a performance measure that characterizes how effectively an evolutionary algorithm locates a given target, given, at most, m steps and given a particular item of information. Next this performance measure is averaged over all the items of information in the informational context. A generic NFL theorem then says that this averaged performance measure is independent of the evolutionary algorithm—in other words, it's the same for all evolutionary algorithms. And since blind search always constitutes a perfectly valid evolutionary algorithm, this means that the average performance of any evolutionary algorithm is no better than blind search.

The significance of the no-free-lunch theorems is that an informational context does not, and indeed cannot, privilege a given target. Instead, an informational context contains information that is equally adept at guiding an evolutionary algorithm to other targets in the phase space. This was certainly the case in Dawkins's METHINKS IT IS LIKE A WEASEL example. The informational context there was the class of all fitness functions, and the information was the fitness function that assigns to each sequence of letters and spaces the number of characters coinciding with METHINKS IT IS LIKE A WEASEL. In the formulation of this fitness function there was nothing special about the sequence METHINKS IT IS LIKE A WEASEL. Any other character sequence of 28 letters and spaces would have served equally well. Given any target sequence whatsoever, we can define a fitness function that assigns the number of places where an arbitrary character sequence agrees with it. Moreover, given this fitness function, the evolutionary algorithm will just as surely converge to the new target as previously it converged to METHINKS IT IS LIKE A WEASEL.

In general, then, there are no privileged targets, and the only thing distinguishing targets is the choice of information from the informational context. But this means that the problem of locating a target has been displaced. The new problem is locating the information needed to locate the target. The informational context thus becomes a new phase space in which we must locate a new target—the new target being the information needed to locate the original target. To say that an evolutionary algorithm has generated specified complexity within the original phase space is therefore really to say that it has borrowed specified complexity from a higher-order phase space, namely, the informational context. And since in practice this new phase space is much bigger and much less tractable than the original phase space, it follows that the evolutionary algorithm has in fact not generated specified complexity at all but merely shifted it around.

Can NFL Be Avoided?

The essential difficulty in generating specified complexity with an evolutionary algorithm can now be put quite simply. An evolutionary algorithm is supposed to find a target within phase space. To do this successfully, however, it needs more information than is available to a blind search. But this additional information is situated within a wider informational context. And locating that additional information within the wider context is no easier than locating the original target within the original phase space. Evolutionary algorithms, therefore, displace the problem of generating specified complexity but do not solve it. I call this the *displacement problem*.

There is no way around the displacement problem. This is not to say that there haven't been attempts to get around it. But invariably we find that when specified complexity seems to be generated for free, it has in fact been front-loaded, smuggled in, or hidden from view. I want, then, in this concluding section to review some attempts to get around the displacement problem and uncover just where the displaced information resides once it goes underground. First off, it should be clear that NFL is perfectly general—it applies to *any* information that might supplement a blind search and not just to fitness functions. Usually the NFL theorems are put in terms of fitness functions over phase spaces. Thus, in the case of biological evolution, one can try to mitigate the force of NFL by arguing that evolution is nonoptimizing. Joseph Culberson, for instance, asks (Culberson 1998, 125), "If fitness is supposed to be increasing, then in what nontrivial way is a widespread species of today more fit than a widespread species of the middle Jurassic?" But NFL theorems can just as well be formulated for informational contexts that do not comprise fitness functions. The challenge facing biological evolution, then, is to avoid the force of NFL when evolutionary algorithms also have access to information other than fitness functions. Merely arguing that evolution is nonoptimizing is therefore not enough. Rather, one must show that finding the information that guides an evolutionary algorithm to a target is substantially easier than finding it through a blind search.

Think of it this way. In trying to locate a target, you can sample no more than m points in phase space. What's more, your problem is sufficiently complex that you will need additional information to find the target. But that information resides in a broader informational context. If searching through that broader informational context is no easier than searching through the original phase space, then you are no better off going with an evolutionary algorithm than with a straight blind search. Moreover, you can't arbitrarily truncate your informational context simply to facilitate your search, for any such truncation will itself be an act of ruling out possibilities, and that by definition means an imposition of novel information, the very thing we're trying to generate.

To resolve the displacement problem, therefore, requires an answer to the following question: How can the informational context be simplified sufficiently so that finding the information needed to locate a target is easier than finding the target using blind search? There's only one way to do this without arbitrarily truncating the informational context, and that's for the phase space itself to constrain the informational context. Structures and regularities of the phase space must by themselves be enough to constrain the selection of points in the phase space and thus facilitate locating the target. The move here, then, is from contingency to necessity; from evolutionary algorithms to dynamical systems; from Darwinian evolution to complex self-organization. Stuart Kauffman's approach to biological complexity epitomizes this move, focusing on autocatalytic reactions that reliably settle into complex behaviors and patterns (Kauffman 1995, chap. 4).

Nonetheless, even this proposed resolution of the displacement problem fails. Yes, structures and regularities of the phase space can simplify the informational context so that finding the information needed to locate a target is easier than finding the target using blind search. But whence these structures and regularities in the first place? Structures and regularities are constraints. And constraints, by their very specificity, could always have been otherwise. A constraint that is not specific is no constraint at all. Constraints are constraints solely in virtue of their specificity—they permit some things and rule out others. But in that case, different constraints could fundamentally alter what is permitted and what is ruled out. Thus, the very structures and regularities that were supposed to eliminate contingency, information, and specified complexity merely invite them back in.

To see this, consider a phase space with an additional structure or regularity that simplifies the informational context, thereby facilitating the task of finding the information needed to locate a target. What sorts of structures or regularities might these be? Are they topological? A topology defines a class of open sets on the phase space. If those open sets in some way privilege the target, then by permuting the underlying points of the phase space, a new topology can be generated that privileges any other target we might choose (that's why mathematicians refer to topology as "point-set topology"). What's more, the totality of topologies associated with a given phase space is vastly more complicated than the original phase space. Searching a phase space by exploiting its topology therefore presupposes identifying a suitable topology within a vast ensemble of topologies. As usual, the displacement problem refuses to go away.

Or suppose instead that the constraint to be exploited for locating a target constitutes a dynamical system, that is, a temporally indexed flow describing how points move about in phase space. Dynamical systems are the stuff of "chaos theory" and underlie all the wonderful fractal images, strange attractors, and self-similar objects that have so captured the public imagination. Now, if the dynamical system in question helps locate a target, it's fair to ask what other dynamical systems the phase space is capable of sustaining and what other targets they are capable of locating. In general, the totality of different possible dynamical systems associated with a phase space will be far more complicated than the original phase space (if, for instance, the phase space is a differentiable manifold, then any vector field induces a differentiable flow whose tangents are the vectors of the vector field; the totality of different possible flows will in this case be immense, with flows going in all conceivable directions). Searching a phase space by exploiting a dynamical system therefore presupposes identifying a suitable dynamical system within a vast ensemble of dynamical systems. As usual, the displacement problem refuses to go away.

Exploiting constraints on a phase space to locate a target is therefore merely another way of displacing information. Not only does it not solve the displacement problem, its applicability is quite limited. Many phase spaces are homogeneous and provide no help in locating targets. Consider, for instance, a phase space comprising all possible character sequences from a fixed alphabet (such phase spaces model not only written texts but also polymers—e.g., DNA, RNA, and proteins). Such phase spaces are perfectly homogeneous, with one character string geometrically interchangeable with the next. Whatever else the constraints on such spaces may be, they provide no help in locating targets. Rather, external semantic information (in the case of written texts) or functional information (in the case of polymers) is needed to locate a target. To sum up, there's no getting around the displacement problem. Any output of specified complexity requires a prior input of specified complexity. In the case of evolutionary algorithms, they can yield specified complexity only if they themselves, as well as the information they employ (typically a fitness function), are carefully adapted to the problem at hand. In other words, all the specified complexity we get out of an evolutionary algorithm has first to be put into its construction and into the information that guides the algorithm. Evolutionary algorithms therefore do not generate or create specified complexity but merely harness already existing specified complexity.

How, then, does one generate specified complexity? There's only one known generator of specified complexity, and that's intelligence.

In every case where we know the causal history underlying an instance of specified complexity, an intelligent agent was involved. Most human artifacts, from Shakespearean sonnets to Dürer woodcuts to Cray supercomputers, are specified and complex. For a signal from outer space to convince astronomers that extraterrestrial life is real, it too will have to be complex and specified, thus indicating that the extraterrestrial is not only alive but also intelligent (hence the search for extraterrestrial *intelligence*—SETI). Thus, to claim that natural laws, even radically new ones, can produce specified complexity is to commit a category mistake. It is to attribute to laws something they are intrinsically incapable of delivering. Indeed, all our evidence points to intelligence as the sole source for specified complexity.

Darwinian Evolution in Nature

Consider carefully what the displacement problem means for Darwinian evolution in nature. Darwinists are unlikely to see the displacement problem as a serious threat to their theory. I've argued evolutionary algorithms like the one in Dawkins's METHINKS-IT-IS-LIKE-A-WEASEL example fail to generate specified complexity because they smuggle it in during construction of the fitness function. Now, if evolutionary algorithms modeled, say, the stitching together of monomers to generate some initial self-replicating polymer, strict Darwinists would admit the relevance of the displacement problem (to paraphrase Theodosius Dobzhansky, to speak of generating an initial replicator via a Darwinian selection mechanism is a contradiction in terms because that very mechanism presupposes replication). Darwinists, however, are principally interested in modeling evolutionary progress once a replicator has come into existence, and here they argue that the displacement problem is irrelevant.

Why? According to Richard Dawkins, nature's criterion for optimization is not an arbitrarily chosen distant target but survival and reproduction, and these are anything but arbitrary. As he puts it in *The Blind Watchmaker*,

> Although the monkey/Shakespeare model is useful for explaining the distinction between single-step selection and cumulative selection, it is misleading in important ways. One of these is that, in each generation of selective "breeding," the mutant "progeny" phrases were judged according to the criterion of resemblance to a distant ideal target, the phrase METHINKS IT IS LIKE A WEASEL. Life isn't like that. Evolution has

no long-term goal. There is no long-distance target, no final perfection to serve as a criterion for selection. . . . In real life, the criterion for selection is always short-term, either simple survival or, more generally, reproductive success. . . . The "watchmaker" that is cumulative natural selection is blind to the future and has no long-term goal. (1986, 60)

The Darwinist therefore objects that "real-life" Darwinian evolution can generate specified complexity without smuggling it in. The fitness function in biological evolution follows directly from differential survival and reproduction, and this, according to the Darwinist, *can* legitimately be viewed as a "free lunch." In biological systems, the replicator (i.e., the living organism) will sample different variants via mutation, and then the fitness function freely bestowed by differential survival and reproduction will select those variants that constitute an improvement, which within Darwinism is defined by being better at surviving and reproducing. No specified complexity is required as input in advance.

If this objection is conceded, then the only way to show that the Darwinian mechanism cannot generate specified complexity is by demonstrating that the gradients of the fitness function induced by differential survival and reproduction are not sufficiently smooth for the Darwinian mechanism to drive large-scale biological evolution. To use another Dawkins metaphor, one must show that there is no gradual way to ascend "Mount Improbable." This is a separate line of argument and one that Michael Behe has developed in *Darwin's Black Box* (Behe 1996). Here, however, I want to show that this concession need not be granted and that the displacement problem does indeed pose a threat to Darwinism.

Things are not nearly as simple as taking differential survival and reproduction as brute givens and from there concluding that the fitness function induced by these is likewise a brute given. Differential survival and reproduction by themselves do not guarantee that anything interesting will happen. Consider, for instance, Sol Spiegelman's work on the evolution of polynucleotides in a replicase environment. Leaving aside that the replicase protein is supplied by the investigator (from a viral genome), as are the activated mononucleotides needed to feed polynucleotide synthesis, the problem here and in experiments like it is the steady attenuation of information over the course of the experiment. As Brian Goodwin notes,

[i]n a classic experiment, Spiegelman in 1967 showed what happens to a molecular replicating system in a test tube, without any cellular organization around it. The replicating molecules (the nucleic acid tem-

plates) require an energy source, building blocks (i.e., nucleotide bases), and an enzyme to help the polymerization process that is involved in self-copying of the templates. Then away it goes, making more copies of the specific nucleotide sequences that define the initial templates. But the interesting result was that these initial templates did not stay the same; they were not accurately copied. They got shorter and shorter until they reached the minimal size compatible with the sequence retaining self-copying properties. And as they got shorter, the copying process went faster. So what happened with natural selection in a test tube: the shorter templates that copied themselves faster became more numerous, while the larger ones were gradually eliminated. This looks like Darwinian evolution in a test tube. But the interesting result was that this evolution went one way: toward greater simplicity. Actual evolution tends to go toward greater complexity, species becoming more elaborate in their structure and behavior, though the process can also go in reverse, toward simplicity. But DNA on its own can go nowhere but toward greater simplicity. In order for the evolution of complexity to occur, DNA has to be within a cellular context; the whole system evolves as a reproducing unit. (1994, 35–36)

My point here is not that Darwinian evolution in a test tube should be regarded as disconfirming evidence for Darwinian evolution in nature. Rather, it is that if the Darwinian mechanism of differential survival and reproduction is what in fact drives full-scale biological evolution in nature, then the fitness function induced by that mechanism has to be very special. Indeed, many prior conditions need to be satisfied for the function to take a form consistent with the Darwinian mechanism being the principal driving force behind biological evolution. Granted, the fitness function induced by differential survival and reproduction in nature is nonarbitrary. But that doesn't make it a free lunch either.

Think of it this way. Suppose we are given a phase space of replicators that replicate according to a Darwinian mechanism of differential survival and reproduction. Suppose this mechanism induces a fitness function. Given just this information, we don't know if evolving this phase space over time will lead to anything interesting. In the case of Spiegelman's experiment, it didn't—Darwinian evolution led to increasingly simpler replicators. In real life, however, Darwinian evolution is said to lead to vast increases in the complexity of replicators, with all cellular organisms tracing their lineage back to a common unicellular ancestor. Let's grant this. The phase space then comprises a vast array of DNA-based, self-replicating cellular organisms, and the Darwinian mechanism of differential survival and reproduction over this phase space induces a fitness function that underwrites full-scale Darwinian evolution. In other words, the fitness function is consistent not only with the descent of all organisms from a common ancestor (i.e., common descent) but also with the Darwinian mechanism accounting for the genealogical interrelatedness of all organisms. Now suppose this is true. What prior conditions have to be satisfied for the fitness function to be the type of fitness function that allows a specifically Darwinian form of evolution to flourish?

For starters, the phase space had better be nonempty, and that presupposes raw materials like carbon, hydrogen, and oxygen. Such raw materials, however, presuppose star formation, and star formation in turn presupposes the fine-tuning of cosmological constants. Thus for the fitness function to be the type of fitness function that allows Darwin's theory to flourish presupposes all the anthropic principles and cosmological fine-tuning that lead many physicists to see design in the universe. Yet even with cosmological fine-tuning in place, many additional conditions need to be satisfied. The phase space of DNA-based self-replicating cellular organisms needs to be housed on a planet that's not too hot and not too cold. It needs a reliable light source. It needs to have a sufficient diversity of minerals and especially metals. It needs to be free from excessive bombardment by meteors. It needs not only water but enough water. Michael Denton's *Nature's Destiny* (1998) is largely devoted to such specifically terrestrial conditions that need to be satisfied if biological evolution on Earth is to stand any chance of success.

But there's more. Cosmology, astrophysics, and geology fail to exhaust the conditions that a fitness function must satisfy if it is to render not just biological evolution but specifically a Darwinian form of it, the grand success we see on planet Earth. Clearly, DNA-based replicators need to be robust in the sense of being able to withstand frequent and harsh environmental insults (this may seem self-evident, but computer simulations with artificial life forms tend to be quite sensitive to unexpected perturbations and thus lack the robustness we see in terrestrial biology). What's more, the DNA copying mechanism of such replicators must be sufficiently reliable to avoid error catastrophes. Barring a high degree of reliability, the replicators will go extinct or wallow interminably at a low state of complexity (basically just enough complexity to avoid the error catastrophe).

But, perhaps most important, the replicators must be able to increase fitness and complexity in tandem. In particular, fitness must not be positively correlated with simplicity. This last requirement may seem easily purchased, but it is not. Stephen Jay Gould, for instance, in

Full House (1996), argues that replication demands a certain minimal level of complexity below which things are dead (i.e., no longer replicate). Darwinian evolution is thus said to constitute a random walk off a reflecting barrier, the barrier constituting a minimal complexity threshold for which increases in complexity always permit survival but decreases below that level entail death. Enormous increases in complexity are thus said to become not only logically possible but also highly probable.

The problem with this argument is that in the context of Darwinian evolution such a reflecting barrier tends also to be an absorbing barrier (i.e., there's a propensity for replicators to stay close to if not right at the minimal complexity threshold). As a consequence, such replicators will over the course of evolution remain simple and never venture into high degrees of complexity. Simplicity by definition always entails a lower cost in raw materials (be they material or computational) than increases in complexity, and so there is a inherent tendency in evolving systems for selection pressures to force such systems toward simplicity (or, as it is sometimes called, *elegance*).

Fitness functions induced by differential survival and reproduction are more naturally inclined to place a premium on simplicity and regard replicators above a certain complexity threshold as too cumbersome to survive and reproduce. The Spiegelman example is a case in point. Thomas Ray's Tierra simulation gave a similar result, showing how selection acting on replicators in a computational environment also tended toward simplicity rather than complexity—unless parameters were set so that selection could favor larger organisms (complexity here corresponds with size). This is not to say that the Darwinian mechanism automatically takes replicating systems toward a minimal level of complexity, but that if it doesn't, then some further conditions need to be satisfied, conditions reflected in the fitness function.

The catalogue of conditions that the fitness function induced by differential survival and reproduction needs to satisfy is vast, if the spectacular diversity of living forms we see on Earth is properly to be attributed to a Darwinian form of evolution. Clearly, such a catalogue is going to require a vast amount of specified complexity, and this specified complexity will be reflected in the fitness function that, as Darwinists rightly note, is nonarbitrary but, as Darwinists are reluctant to accept, is also not a free lunch. Throw together some replicators, subject them to differential survival and reproduction, perhaps add a little game theory to the mix (à la Robert Wright), and there's no reason to think you'll get anything interesting, and certainly not a form of Darwinian evolution that's worth spilling any ink over. Thus I submit that even if Darwinian evolution is the means by which the panoply of life on Earth came to be, the underlying fitness function that constrains biological evolution would not be a free lunch and not a brute given but a finely crafted assemblage of smooth gradients that presupposes much prior specified complexity.

REFERENCES

Behe, Michael. 1996. *Darwin's Black Box: The Biochemical Challenge to Evolution*. New York: Simon and Schuster.

Culberson, Joseph C. 1998. "On the Futility of Blind Search: An Algorithmic View of 'No Free Lunch.'" *Evolutionary Computation* 6:2.

Davies, Paul. 1999. *The Fifth Miracle. The Search for the Origin and Meaning of Life*. New York: Simon and Schuster.

Dawkins, Richard. 1986. *The Blind Watchmaker*. New York: Norton.

Denton, Michael. 1998. *Nature's Destiny: How the Laws of Biology Reveal Purpose in the Universe*. New York: Free Press.

Dretske, Fred. 1981. *Knowledge and the Flow of Information*. Cambridge, Mass.: MIT Press.

Eigen, Manfred. 1992. *Steps Towards Life: A Perspective on Evolution*. Translated by P. Wooley. Oxford: Oxford University Press.

Goodwin, Brian. 1994. *How the Leopard Changed Its Spots: The Evolution of Complexity*. New York: Scribner's.

Gould, Stephen Jay. 1996. *Full House: The Spread of Excellence from Plato to Darwin*. New York: Harmony Books.

Kauffman, Stuart. 1995. *At Home in the Universe: The Search for the Laws of Self-Organization and Complexity*. New York: Oxford University Press.

Stalnaker, Robert. 1984. *Inquiry*. Cambridge, Mass.: MIT Press.

SEVEN

THE SECOND LAW OF GRAVITICS AND THE FOURTH LAW OF THERMODYNAMICS

Ian Stewart

There is a fundamental difference between galaxies and systems that are normally dealt with in statistical mechanics, such as molecules in a box. This difference lies in the nature of the forces that act between the constituent particles. The force between two gas molecules is small unless the molecules are very close to each other, when they repel each other strongly. Consequently gas molecules are subject to violent and short-lived accelerations as they collide with one another, interspersed with longer periods when they move at nearly constant velocity. The gravitational force that acts between the stars of a galaxy is of an entirely different nature . . . the net gravitational force acting on a star in a galaxy is determined by the gross structure of the galaxy rather than by whether the star happens to lie close to some other star.

J. Binney and S. Tremaine, *Galactic Dynamics*

There are several ways of defining and interpreting statistical entropy. The arbitrariness is connected with both the assumed probability field and with the nature of constraints used in the coarse-graining process. Both Boltzmann and Gibbs were aware of these ambiguities in the statistical interpretation of thermodynamic variables.

J. M. Jauch and J. G. Barón, *Entropy, Information, and Szilard's Paradox*

Physics provides a number of universal principles that constrain the possible behavior of real-world systems, of which the first, best established, and most fundamental is the law of conservation of energy in classical mechanics. Similar conservation laws include conservation of linear and angular momentum. Later, the laws of thermodynamics were added, among which the best known and traditionally the most puzzling is the second law of thermodynamics. This associates with any *closed* thermodynamic system a quantity known as *entropy* (normally interpreted as a measure of disorder) and says that as time passes, the total entropy of the system always increases. Despite early inferences about the "heat death of the universe," Dyson (1979) showed that life-like behavior can continue indefinitely even as entropy increases. Thanks to the seminal work of Shannon and Weaver (1964), a third quantity has joined the energy/entropy duo: *information*. Information is often interpreted as the opposite of entropy (arguably wrongly: see appendix 7.1) and is viewed as a source of organization; this in turn is used to explain *complexity*. A key example: The complexity of a living organism is often attributed to the information contained in its DNA, as, for example, in Dawkins (1987). Some profound physics has come out of these conceptual developments, especially when allied to principles of general relativity (big bang, black holes) and quantum mechanics: A clear example is "black hole thermodynamics" (Hawking 1988). Relations between information and irreversible processes have been developed (see Landauer 1961; Bennett 1973, 1982). Many physicists wonder whether there is some law of conservation of information, in which case all of the universe's information store, hence (they assert) all of its complexity, is limited by the amount of information that was put into it in the big bang. (However, if you believe in conservation of information, then you can't also believe that information is negative entropy *and* that the second law of thermodynamics is valid.) And some physicists argue that the universe is *made* of information (see Toffoli 1982)—just as we used to think it was made from energy, machinery, or the mutilated corpse of a murdered goddess (as in the Babylonian epic *Enuma Elish*).

Unlike the universe of the physicists, the biological world (henceforth "biosphere") gives the appearance of progressing toward increasing complexity and more sophisticated organization—the exact opposite of the second law of thermodynamics. There is no logical discrepancy here, because the biosphere is driven by energy inputs—mainly from solar radiation—and hence is not a closed system. Nevertheless, there is a *phenomeno*logical discrepancy: Physics provides no obvious reason why a thermodynamic system provided with an energy input should tend toward greater organization. On the contrary, it provides numerous examples where this *fails* to occur, such as water boiling and becoming disordered steam. So lack of closure explains why the second law of thermodynamics is inapplicable, but it doesn't explain what actually happens. To plug this explanatory gap, many authors, notably Kauffman (1993, 1995, 2000), have speculated about a "fourth law of thermodynamics"—some general physical principle supplementing the standard three laws of thermodynamics—that explains *why* the biosphere tends, in some sense, toward increasing complexity. (In particular, Kauffman 2000 formalizes a living organism or life-like system as an "autonomous agent" and conjectures that autonomous agents expand into the "space of the adjacent possible" at a maximal rate.) Others, including many evolutionary biologists, have argued that there is no evidence for growing complexity in the biosphere anyway. This thesis can be argued quite persuasively, although one does wonder what circumstances brought about the human ability to construct persuasive arguments.

The biosphere is not alone as a source of increasing complexity. Indeed Davies (1992; this volume, chapter 5) points out that the most convincing real-world example of increasing complexity is a purely physical one: the growing clumpiness of the universe. With the passage of time, an initially smooth universe has formed itself into bigger and bigger clumps—for example, the "Great Wall" is some 500 million light years across (Powell 1994).

When thinking about these issues it helps to keep the conceptual issues simple, and this can be done by considering the universe as a self-gravitating (henceforth "gravitic") system of particles and ignoring all other physical features. The issue of increasing clumpiness is already apparent in a simple model in which the universe is considered as a system of (very many) identical spheres that obey Newton's law of gravitation. Mathematically, such a system has an appealing conceptual simplicity and clarity, comparable to that of the "hard sphere" model of a thermodynamic gas. These two simplified models illuminate many of the key issues of entropy and complexity. Here, I use them to argue that on the level of description normally employed in thermodynamics, it is in the nature of gravitic systems to run in the opposite direction to thermodynamic ones—that is, from "disorder" to "order," not from "order" to "disorder." (Here I use these two terms loosely, as is customary. Whether this is a good idea is a separate issue—see appendix 7.2.) This is not a new idea, as this chapter's epigraphs make clear, but it seems repeatedly to be forgotten.

At first sight these observations hold out hope for a "second law of gravitics" that quantifies the increasing complexity of gravitic systems in a similar manner to the second law of thermodynamics—though the technical problems in formulating such a law are considerable. But there is a deeper problem, which relates not to mathematical technicalities but to modeling assumptions.

All mathematical models of the physical world involve approximations and simplifications and are therefore valid only within certain limitations—such as spatial or temporal scales and statistical fluctuations. My main argument here is that *the modeling assumptions appropriate to a second law of gravitics conflict with those traditionally assumed in thermodynamics*. The domains of validity of the thermodynamic and gravitic modeling assumptions overlap quite well for some purposes, but they appear not to overlap well enough for a hypothetical second law of gravitics simply to be added to the laws of thermodynamics.

Moreover, we can consider "second law of gravitics" a place-holder for the hoped-for fourth law of thermodynamics, so the same goes for the fourth law. *Whatever it might be, it cannot be a law of thermodynamics in the same general sense as the first three laws, and it cannot be merely adjoined to them.* This does not rule out *some* kind of fourth law, but it has to be a context-dependent law. Encouragingly, Kauffman's speculations about the adjacent possible have context built in—indeed, "space of the adjacent possible" is a formal specification of the local context of an autonomous agent.

Time-Reversal Symmetry of Classical Mechanics

It is well known that Newtonian mechanics is time-reversible. It is worth setting this idea up precisely, in order to avoid later misconceptions. Throughout this chapter I will ignore issues related to quantum mechanics and general relativity, adopting a purely classical view of mechanics. This is done to simplify the discussion and highlight key issues, but I don't want to give the impression that these two physical paradigms are irrelevant. Thermodynamics, in particular, is now routinely derived in a quantum context. However, a lot of the confusion in the area can be sorted out using classical physics, and my general line of argument could probably be reformulated in a quantum or relativistic context.

In classical mechanics (as elsewhere) time-reversibility follows from a simple mathematical feature of Newton's laws of motion: time-reversibility is a *symmetry* of classical Newtonian mechanics (provided that the mass of the particle, and the forces that act, are time-independent). Recall that Newton's main law of motion for a particle says that acceleration is proportional to force, and takes the mathematical form

$$F = mx'' \qquad (1)$$

where $x = x(t)$ is the position of the particle at time t and $''$ indicates the second time derivative. Substituting $-t$ for t we find that $x'(-t) = -x'(t)$, so that $x''(-t) = x''(t)$. So if $x(t)$ is a solution of the Newtonian equation (1), and F and m are constant, then $x(-t)$ is also a solution. The same goes for more complex configurations of many particles and many rigid bodies. Intuitively, if you make a movie of feasible Newtonian dynamics and run it backward, you get another movie of feasible Newtonian dynamics. In contrast, "Aristotelian" dynamics—a simple, approximate model of Aristotle's ideas that corresponds to a world with substantial friction—can be described by the equation

$$F = mx' \qquad (2)$$

in which *velocity* is proportional to force. Let the force decline to zero, and the motion stops. Changing t to $-t$ does not preserve this equation: in fact it turns it into $F = -mx'$. So Aristotelian dynamics has a uniquely defined time direction, the direction in which friction causes motion to die away.

Although Newton's equations are time-reversible, *solutions* of Newton's equations—trajectories of a system that obeys them—need not be (Baez 1994; Zeh 1992). (Indeed, in a reasonable sense, "most" are not.) This is an example of the general phenomenon known as *symmetry-breaking*, which arises because individual solutions of a system of equations can (and often do) have *less* symmetry than the equations themselves (for an extensive discussion in nontechnical language, see Stewart and Golubitsky 1992). In fact, a time-reversibly symmetric *solution* of Newton's equations would be very strange: the history of the entire universe would "bounce" at time zero and thereafter follow its previous trajectory, but in reverse. (Simple mechanical systems often have such solutions. As an example: consider a ball tossed vertically into the air in a uniform gravitational field. It rises, slows, and then turns and descends. If we take the origin of time to occur when it is at the apex of its trajectory, the whole motion is time-reversible.) Although such solutions are in principle possible, you wouldn't expect to live in one. For most solutions, time-reversal yields a *different* solution. This means that we can inhabit a universe with a clear arrow of time, even though there might be a dual universe whose arrow of time reverses the one observed in ours.

In summary: one way to exhibit a time-asymmetric solution of a time-symmetric equation is to specify time-asymmetric *initial conditions*. Having started your universe out in a time-asymmetric manner, you shouldn't be surprised if it continues to behave in a time-asymmetric manner. In particular, the time-reversibility of Newtonian mechanics does not imply anything about the behavior of entropy in some particular realization of the universe's dynamics. It could increase, decrease, stay constant, or even wobble up and down.

Thermodynamics

Thermodynamics began as a description of gases and other bulk molecular systems in terms of macroscopic variables such as pressure, density, and volume. Its main thrust was to codify the relation between mechanical energy and heat, and its derivation relied heavily on the metaphor of heat engines. The second law of thermodynamics, which says that the entropy of a closed thermodynamic system increases with the passage of time, successfully rules out perpetual motion machines "of the second kind," which employ heat energy as well as mechanical energy. Later, thermodynamics was reformulated by Boltzmann (1909) and others, who based it on a statistical description of ensembles of discrete molecules (for a monatomic gas, they are taken to be spheres) obeying classical mechanics in a simplified context (perfectly elastic collisions). At this point a conflict between the two types of model was recognized, especially by Loschmidt (1869): the thermodynamic principle of increasing entropy is inconsistent with the time-reversibility of classical mechanics. The paradox arises because of an apparent conflict between two approaches to the dynamics of a gas: a "continuum" model that views the gas as a continuous distribution of infinitely divisible matter (for which entropy increases) and a "discrete" model that views it as a collection of tiny hard spheres (for which time-reversibility holds). Boltzmann explained the inconsistency in two (related) ways:

- Thermodynamic laws are statistical in character: they are valid with (very) high probability, but they are not universal.
- The evolution of the universe depends on its initial conditions, and the choice of these breaks the time-reversal symmetry of the laws of motion.

Both of these observations are correct, but even taken together they do not provide a satisfactory resolution of Loschmidt's "paradox." To see why, it is necessary to recall a few pieces of background on thermodynamics.

As just said, there are two models: a continuum one and a discrete one. In the continuum model, which historically came first, a gas is described by quantitites such as pressure, temperature, and density. These are assumed to vary *continuously* across some spatial region and to remain *constant* under statistical fluctuations around equilibrium states. Such a description—if taken literally—becomes invalid on smaller scales than the average gap between gas molecules, but for many purposes (especially heat engines) it works well. Nevertheless, some of the apparent paradoxes in the relation between classical mechanics and thermodynamics can be traced to the modeling assumptions of the continuum approximation, as I'll show.

A later alternative, statistical mechanics, introduced what is in a sense a "hidden variable" underpinning for continuum thermodynamics. This is the standard microscopic model of a thermodynamic gas: a large number of identical "hard" spheres. These obey Newton's laws of motion, augmented by perfectly elastic bouncing whenever two spheres collide. It is assumed that no other forces (such as gravity) are acting. The "force law" is thus an extreme case of a short-range repelling force—short-range because it is zero unless spheres come into contact, and repelling since when they do they subsequently move apart. A system in "thermodynamic equilibrium" is *not* in mechanical equilibrium on the microscopic level. On the contrary, it is abuzz with motion—molecules whizzing this way and that, colliding, bouncing, and zigzagging almost at random. Only the statistical properties of the system, such as density, are constant—and then only approximately, though the "error" is *tiny*.

Consider, for example, the concept "density." Assume that some region of space (the *domain* of the model) contains a large number of identical small spheres. Subdivide that region into boxes that are small by comparison with the domain but still contain large numbers of spheres. Then the density $D(B)$ of the gas inside a given box B, at a given instant of time, can be quantified in units of "molecules per unit volume"—that is, by the number $N(B)$ of spheres that the box contains, divided by the volume $V(B)$ of the box. Mathematically,

$$D(B) = N(B)/V(B).$$

We say that the boxes constitute a *coarse-graining* of the domain: this is the first modeling approximation.

If a sphere enters or leaves B, which is a common occurrence, then the density $D(B)$ increases or decreases by an amount $1/V(B)$. That is, density can assume only a discrete set of values (unless we count parts of a sphere). However, if $N(B)$ is large, such a change is negligible in proportion to the whole. It is therefore a reasonable approximation to assume that $D(B)$ takes values on a continuous scale of measurement. For a system that is in "equilibrium," the value of $D(B)$ remains constant for each box B, give or take an error that is small enough to be negligible. So already we have two more approximations (closely related to each other). There is a third, also closely related, which is to replace the set of boxes by continuous spatial coordinates, and a fourth, yet again closely related, which is to modify the values of $D(B)$ so that density becomes a continuous variable. For example, we might associate B with its central point X_B, define the local density *at the point X_B* to be $D(B)$, and then smoothly interpolate to get a quantity $D(X)$ defined at every point X in the domain. This interlinked set of approximations is called *smoothing*, and it converts the statistical quantities defined by coarse-graining into well-defined continuous functions amenable to the analytic methods of the calculus. In most textbook treatments, all of the preceding is tacit.

In conjunction, these approximations—coarse-grain and then smooth —are so good that they provide an excellent physical description of the instantaneous state of a gas, to a level of accuracy that is far better than any experiment. They also provide good descriptions of how such a state changes with time, over time-scales longer than those of any practicable experiment. To many physicists, this means that the model is simply "true," and so it is for many purposes—but when one is discussing conceptual issues such as the arrow of time and trying to play off the microscopic mechanical description against the macroscopic continuum one, one cannot ignore the approximations: they are of the essence.

The Thermodynamic Arrow of Time

It is often said that thermodynamics provides an "arrow of time." Its role in this respect is often overstated, though: Thermodynamics does not explain why time exists—that is, it does not *give rise* to a time variable. The issue is clearer in the discrete model, which *assumes* a "line of time"—an explicit one-dimensional time variable t in the equations. What entropy does (without full logical rigor but with a good correspondence to certain features of the physical universe) is to attach a specific arrow to this line, when in principle the arrow might point either way. The discrete model is compatible with both ways to place the arrow, thanks to time-reversibility; the continuum model, with entropy increase built in, chooses one of these (but only for systems where entropy either always increases or always decreases). The issue, then, is one of choice between two directions of time-flow, in the presence of very specific assumptions about the nature of time and its physical role; not the "construction" of time itself.

Thermodynamics describes the state of a gas by a small list of continuous variables. I have already mentioned temperature, density, volume, and pressure; to these must now be added the rather subtle concept of *entropy* (there are also others, such as enthalpy, which I shall ignore). Formally, the difference in entropy between two states is defined to be the integral (along any "reversible" path) of q/T, where q is the heat being received and T is temperature. Reversibility here has a technical definition, but its general import is that the change happens so slowly that the system remains in (actually, very close to) equilibrium throughout, and any heat flow that occurs can be reversed by running the change backward.

Entropy is often described informally as "disorder," an image that is useful provided you bear in mind where it came from. We have a simple mental image of "order" versus "disorder"—clothes neatly pegged on a washing-line instead of lying all over the floor, bricks laid to form a wall rather than a messy heap. Both chaos theory and complexity theory have taught us that the order/disorder metaphor is fraught with tacit assumptions (that are often wrong) and is inconsistent, ill defined, and by no means unique (see appendix 7.2). Moreover, it is highly context-dependent (see Cohen & Stewart 1994, 254–257, for an example). The formal definition—asserted within the thermodynamic model and therefore carrying all of the same contextual baggage of coarse-graining and smoothing—allows us to calculate the entropy of an equilibrium state of a gas, and this is where the order/disorder interpretation seems to come from. It goes like this. Imagine a gas enclosed in a container, and suppose that momentarily the gas happens to occupy only one corner of the container. (For example, it might temporarily be held there by an impermeable membrane.) Such a situation does not persist: According to thermodynamics, the gas will rapidly distribute itself uniformly throughout the whole of the container. It can be calculated that such a change leads to an increase of entropy. Now: the initial state, with the gas confined to a small corner, has elements of "order"—the molecules have been "tidied away." By comparison, the final state is more disordered. *Ergo*: loss of order corresponds to gain of entropy, so entropy equals disorder.

Fair enough—but there may be equally reasonable examples of "order" that do not work in the same way. Roughly speaking, if order is defined in terms of macroscopic variables the metaphor is a good one, but it may not be if the definition involves microscopic variables. Imagine, for example, a gas whose state is perfectly uniform (thus fully disordered) but *will*, in one second's time, result in the gas occupying only one corner of the enclosure. (Such states exist: start from a state in which all the molecules are in one corner and run it backward in time for a second. For most initial states of this kind, the resulting state will occupy the entire enclosure uniformly—thanks to the second law! Now run forward: the state spontaneously clumps.) Such states surely deserve to be considered "ordered"—just watch, and you'll quickly see why—but their order is not capturable by instantaneous thermodynamic state variables. This is an example of the "implicate order" of Bohm (1980), and it has parallels with the algorithmic information theory of Chaitin (1987) since it considers not just states but *processes* that generate states. Notice that in this example entropy *decreases* instead of increasing.

Phase Space

The discussion that follows will make more sense if I introduce a few useful items of mathematical imagery. These all relate to a geometric formulation of classical mechanics that is normally credited to Henri Poincaré in the late 1890s. The idea has an extensive "prehistory" in the work of William Rowan Hamilton and Joseph Liouville, among others.

In this formulation, at any instant of time a dynamical system exists in some particular *state*, which can be specified by assigning values to a system of *state variables*. Because Newton's laws of motion depend on *second* derivatives (acceleration x''), a system's future motion depends on both the positions of its constituent bodies *and* their velocities (momenta). We can think of these positions and momenta as determining a system of coordinates on some *phase space*. This term seems to have originated with Poincaré, and it is unclear why the word "phase" was employed: an alternative is *state space*. The space of positions (alone) is called *configuration space*. A point in phase space specifies the dynamical state of the entire system (read off its coordinates). As time passes, those coordinates can change, so the point in phase space can move. The laws of motion prescribe how it moves. For any given initial data (point in phase space), the passage of time carries that point along a unique curve in phase space, called its *trajectory*. By choosing different sets of initial data we obtain other trajectories; the set of all trajectories defines a *flow* on the phase space, analogous to the streamlines of a flowing fluid. The future (or past) of a given initial state can be found by following the trajectory through it in either the forward-time or backward-time direction—that is, by letting the initial point follow the flow-lines. Similarly the future (or past) of a given *set* of initial states can be found by following the trajectories through that set in either the forward-time or backward-time direction.

Liouville showed that this flow corresponds to that of an *incompressible* fluid. That is, there exists a notion of "volume" with the property that as the fluid flows along trajectories, the volume of any region of the fluid does not change. This notion of volume is known as an *invariant measure*, and it can be thought of as representing the probability (in some intrinsic sense) of finding the system's state within a given region of phase space. Depending on the system (and the initial data), the flow on phase space may be "turbulent" or "laminar," meaning that it either mixes the "fluid" up a lot, or it doesn't. Roughly speaking, these types of flow correspond to chaotic and nonchaotic dynamics, respectively. The most extreme form of nonchaotic dynamics—that is, the most regular and predictable kind—is called "completely integrable," and the term "laminar" ought perhaps to be reserved for that. The flow may be turbulent in some parts of phase space and laminar in others. For the pendulum, the flow is laminar everywhere.

Boltzmann and Statistical Mechanics

Boltzmann (1909) clearly had a good understanding of the relation between the discrete "bouncing spheres" model and the continuum model of classical thermodynamics, and he found (semirigorous) links between them that enabled him to define entropy in a different and illuminating way. Ignore the smoothing step in the thermodynamic formalism and consider, as earlier, a large number of small boxes, each containing a large number of spheres. The boxes should divide up phase space—positions and velocities—and not just configuration space. Each box has a single *macrostate* determined by statistical averages (pressure, temperature, etc.), and as far as continuum thermodynamics is concerned, the actual disposition of molecules within the box is otherwise irrelevant. (There is no practical way to measure it anyway, but let us keep practical and conceptual issues separate.) Call this disposition of molecules the *microstate* of the box. Coarse-graining has the effect of lumping together all microstates that lead to the same macrostate. Boltzmann defined the entropy of a macrostate to be proportional to the *logarithm* of the number of compatible microstates. Auyang describes how Boltzmann's idea works for gas that is initially confined to half of a box and subsequently spreads into the whole box (Auyang 1998, 319). Consider N spheres in a box. The state of each sphere is determined by three position coordinates and three velocity coordinates, six in all. The entire system requires N sets of six coordinates, one for each sphere, leading to a $6N$-dimensional *microstate* phase space for the whole system. Now we imagine partitioning the microstate phase space into cells—for example, by specifying a range of positions and a range of velocities for each sphere. A given macrostate of the gas will in general correspond to many different cells in microstate phase space—this is what "coarse-grain" means.

Consider two different macrostates: X, a gas that occupies only one-half of the box, and Y, a gas that occupies the entire box. Let x be the region of microstate phase space corresponding to X, and let y be the region of microstate phase space corresponding to Y. Since twice as many positions in microstate space are available to each of the N spheres if it can range over the entire box instead of just half, the number of microstates in y is 2^N times the number of microstates in x. Therefore

$$\text{entropy of } Y = \text{entropy of } X + N \log 2.$$

Since in practice N is of the order of 10^{24}, we see that Y has enormously greater entropy than X. The reason is that the fraction of microstate space corresponding to X is negligible in comparison with that corresponding to Y.

Making this approach rigorous is not easy (see the second epigraph). For example, the number of microstates compatible with a given macrostate is actually infinite. Therefore, we must either perform a second coarse-graining of microstate phase space or use the volume of a region in place of the number of microstates it contains. Moreover, the mathematician's prejudice is that "volume" should be calculated not in terms of naive boxes but in terms of an invariant measure. Nonetheless, in some sense Boltzmann had a good idea. He was even able to demonstrate the second law of thermodynamics within his framework: In a closed system, entropy in the Boltzmann sense (almost) always increases. However, Boltzmann's deduction assumes (without proof) one further principle, which he called "molecular chaos." It says that the probability distribution for the velocities of one molecule is independent (in the probabilistic sense) from the probability distribution for the velocities of any other molecule. This amounts to assuming the "ergodic hypothesis," which can be interpreted as stating that both the positions and the velocities of molecules smear out smoothly as time passes.

Poincaré Recurrence

The "paradox" of entropy increase is sometimes explained (away?) by the phenomenon of *Poincaré recurrence*: if you set up some generic nonequilibrium initial state of a gas and then wait long enough, then the gas will repeatedly return as closely as you wish to that initial state. If the initial state is "ordered"—whatever that means—then it keeps returning to an ordered state. This behavior obviously conflicts with entropy-increase.

The relation between Poincaré recurrence and the second law of thermodynamics tells us something about the coarse-graining process. Suppose we accept that with probability incredibly close to 1, say $P = 1 - 10^{-100}$, the entropy of any given macrostate will increase throughout the next million years. Start with some fixed microstate X_0 and the corresponding macrostate $X_0^\star$. After a million years we reach microstate X_1 and macrostate $X_1^\star$. After k million years we reach microstate X_k and macrostate $X_k^\star$. Poincaré recurrence tells us that for large enough k, we have X_k very close indeed to X_0, so somehow entropy has to *decrease* at some stage. The point is that this is entirely consistent with the statistical statement that "with overwhelming probability entropy always increases." The probability that entropy increases throughout all k transitions is $P^k = (1 - 10^{-100})^k$. Even though P is very close to 1, the value of P^k tends toward *zero* as k increases. If we take $k = 2.10^{101}$, for example, it is less than 0.000001.

In summary: The long-term microdynamics conspires to create states that in the short-to-medium term are highly improbable. If you keep throwing dice long enough, eventually you will get a million sixes in a row. However, the Poincaré recurrence argument is a bit of a red herring, since for any realistic gas the time taken for a given state to recur is considerably longer than the lifetime of the universe. Here I have contemplated a period of 2.10^{107} years, for instance, but it is doubtful whether the universe will last more than 10^{60} years.

Loschmidt and Reversibility

The most enthusiastic proponent of time-reversibility as an argument *against* the second law of thermodynamics was Josef Loschmidt, in 1869. Loschmidt disliked the second law's implication that the long-term fate of the universe was "heat-death," a lukewarm, uniform, totally disordered, maximum-entropy soup. He devised examples to show that whether or not the second law was a valid description of *most* real situations, there were cases where it failed. He derived a counterexample, as follows. Consider a system of particles that are all at rest at the bottom of a container, except for *one* particle that begins some distance away from them and is moving toward them. It collides with the others, causing them to start moving, and (with high probability) the ensuing motion leads to a fully disordered, equilibrium state in which the particles occupy the entire container.

Take this state and preserve each particle's position but exactly reverse its velocity. Now you have a state whose evolution in time is the precise reverse of the first setup. Macroscopically it is in equilibrium. But allow time to run, and (the reversal of) the initial situation will arise: all particles at rest at the bottom of a container, except for *one* particle that is moving away from them.

Boltzmann quickly realized that Loschmidt was correct, in the sense that Boltzmann's theory was incompatible with the time-reversibility of Newtonian mechanics. He concluded, correctly, that there is no completely rigorous way to define entropy in terms of a mechanical micromodel without introducing some assumption that leads to a preferred arrow of time. He resolved the paradox (or claimed to) by appealing to the fact that thermodynamics has only a *statistical* validity. Disordered (high-entropy) states, he claimed, are far more probable than ordered (low-entropy) ones (as seems to be the case in Auyang's example, and Loschmidt's), so with overwhelming probability the state tends from order toward disorder. This led him to the definition of entropy in terms of numbers of microstates: Loschmidt's objection engendered Boltzmann's triumph.

There is, however, a problem, which Boltzmann failed to notice. Essentially it is that the time-reversal of a given state (keep all positions but reverse all velocities) has the same number of microstates as the state itself. And the fact that the second law is only statistical does not help, for statistically any given state ought to have the same probability as its time-reversal. Let me be more precise. We can define the entropy of a *microstate* to be equal to that of the corresponding macrostate—the entropy that we would observe if we were to coarse-grain and look at the macrostate of the system. Let me call this *microentropy*. Mathematically this definition is precise, but it has some awkward properties. Specifically, the evolution in time of microentropy is not uniquely defined, and microentropy need not increase over time. The reason is the loss of precision involved in coarse-graining. Once again, take an "ordered" microstate X that evolves into a "disordered" one Y. Time-reverse Y (flip all velocities) to get $Y^\star$: now $Y^\star$ evolves into X in forward time. However, one expects (according to Boltzmann) "almost all" microstates with the same macrostate as $Y^\star$ to become more disordered, or at least not to become less disordered. So let Z be a microstate with the same macrostate as $Y^\star$, so that at the time t when $Y^\star$ evolves into X, Z evolves into a disordered state W. Now $Y^\star$ and Z have the same microentropy, but they evolve, respectively, into X and W, which have different microentropies. Moreover, in the transition $Y^\star \to X$ the microentropy *decreases*.

Microstates can now be classified according to how their microentropy varies with (forward and backward) time. Here is the start of such a classification:

- Type C: microentropy remains constant
- Type I: microentropy increases monotonically
- Type D: microentropy decreases monotonically
- Type DI: microentropy increases monotonically until some instant and thereafter decreases monotonically
- Type ID: microentropy decreases monotonically until some instant and thereafter increases monotonically

This can be followed by more complex alternations of entropy increase and decrease—and conceivably worse behavior still. The classification could be refined to include not just the direction of change of entropy but its actual value at any given time.

Under time-reversal, type C states remain type C (and are thermodynamically uninteresting since entropy is constant). Type I states become type D and conversely. Type DI remains type DI, and type ID remains type ID. (The rule is: Reverse the symbol sequence and interchange all Is and Ds.) Next, I ask: What is the probability of a given microstate being of a given type? The answer depends on the choice of probability measure. On the microlevel it is reasonable to require this measure to be preserved by time-reversal (it is for the Liouville measure); if so, then *whatever* the chosen measure, the chance of having a type I state is exactly the same as that of having a type D state. If the measure does not satisfy this time-reversibility requirement, then an explicit time-asymmetry has been introduced, and the question of time's arrow ceases to be puzzling. Now Boltzmann's conception of the probability measure does indeed introduce such a time asymmetry, because he defines the measure naïvely in terms of counting boxes in a coarse-grained picture. So Boltzmann's belief that entropy-increasing states are common, whereas entropy-decreasing states are rare is—yet again—an artefact of coarse-graining. To him, "common" and "rare" refer to the coarse-grained world of macrostates, not to the world of microstates.

To a mathematician there are other questionable features of Boltzmann's approach, too: for example, the "number" of microstates (and indeed of macrostates) is infinite. We can get around this by using the volume of relevant parts of suitable phase spaces instead, but now we must ask: "Volume in what sense?" There is no good reason for employing a naïve decomposition into equal-sized cells. Mathematics strongly suggests using a notion of volume ("measure") that has an intrinsic connection to the dynamics, that is, an invariant measure. But now boxes change shape as they propagate dynamically—and while the whole point of such a measure is that the *volumes* of the boxes do not change as they propagate, chaotic dynamics implies that a single initial box can wind itself thinly through a large number of other boxes as time passes, getting tangled up like spaghetti. This points to difficulties in making sense of "microstate" with respect to any *fixed* system of boxes. The general implication is that such an approach works *only* when we also coarse-grain the system. So, like it or not, coarse-graining seems to be an unavoidable feature of Boltzmann's way of thinking.

Gravitational Clumping

Having established the central role of coarse-graining in the second law of thermodynamics, I move on to gravitic systems and examine how coarse-graining works in this case. As I remarked at the beginning, Davies (1992; this volume, chapter 5) has pointed out that the most convincing case of increasing complexity in the universe may well be the universe itself. The matter in the universe started out in the big bang with a very smooth distribution and has become more and more clumpy with the passage of time. Matter is now segregated on a huge range of scales—into rocks, asteroids, planets, stars, galaxies, galactic clusters, galactic superclusters, and so on. Using the same metaphor as in thermodynamics, the distribution of matter in the universe seems to be becoming increasingly ordered, whereas a thermodynamic system should become increasingly disordered. And the universe is, among other things, a thermodynamic system. So gravitic systems are an excellent test-bed for a fourth law of thermodynamics. If we can't formulate such a law for gravitics, what price evolutionary biology?

The cause of this clumping seems to be well established: it is gravity. Smooth distributions of self-gravitating matter are *unstable*: arbitrarily small fluctuations can cause them to become nonuniform. Until recently it looked as though gravitational instabilities alone could not account for the rate at which the universe became clumped—COBE satellite data showed that the initial clumpiness soon after the big bang was very small, and the rate at which gravity can amplify inhomogeneities seemed too slow. Apparently this discrepancy has disappeared in more recent analyses, so I am free to assume, as Davies does, that gravity causes the clumpiness. As we shall see, this makes good sense, and even if it is not the complete picture, it leads to a clear mathematical distinction between gravitic and thermodynamic systems.

Taking at face value the observation that the universe does indeed become increasingly clumpy as time passes, we are faced with an apparent paradox that parallels the paradox of entropy-increase in a time-reversible universe. In either a Newtonian or an Einsteinian model of gravitation, the equations for gravitational systems (Newton's laws of motion plus inverse square law gravitational forces or Einstein's field equations, respectively) are time-reversible. This means, as before, that if any solution of the equations is time-reversed then it becomes an equally valid solution. Our own universe, run backward in this manner, then becomes a gravitational system that gets less and less clumpy as time passes.

The argument of Paul Davies runs as follows: As with all arrows of time, there is a puzzle about where the asymmetry comes in. Therefore *the asymmetry must be traced to initial conditions*. The puzzle then deepens, because we know that the initial conditions were extremely smooth, whereas the "natural" state for gravitational systems presumably should be clumped. It then seems that the initial conditions of the universe must have been very special (an attractive proposition for those who believe that our universe is highly unusual, and ditto for our place within it). Davies reports work of Penrose that quantifies the specialness of this initial state by comparing the thermodynamic entropy of the initial state with that of a hypothetical (but plausible) final state in which the universe has become a system of black holes. This is an extreme degree of clumpiness—though not the ultimate, which would be a single giant black hole—and has the virtue that its entropy is well defined and computable, in the sense made explicit by Hawking. (The entropy of a black hole is currently defined to be proportional to its surface area, a notion that Hawking strongly resisted until he understood in what sense a black hole can have a temperature: see Hawking 1988.) The result is that the initial state has an entropy about 10^{-30} times that of the hypothetical final state—special indeed. This discrepancy leads Penrose to postulate a new time-asymmetric law forcing the early universe to be smooth.

I will argue that this entire exercise is misguided and that the source of the confusion is a mirror image of the usual confusion about time-reversibility in thermodynamic systems. In particular, initial conditions are *not* the only way to account for time-asymmetry in a time-reversible system, as we have already seen: a far more robust source is coarse-graining.

Are Initial Conditions the Answer?

Mathematically, a system of hard spheres with short-range "bounce" forces is—just about—tractable. This tractability stems from the feature that particles move independently *except* for occasional collisions—and that even collisions preserve certain relationships between velocities. The mathematics was tractable enough for Boltzmann to *deduce* the usual laws of continuum thermodynamics from it, along with the "Maxwell distribution" for the probability of a given molecule having a given velocity.

Self-gravitating systems are very different. The long-range nature of the force means that spheres are *never* independent of each other. If one moves, then the force it exerts on any other sphere changes, and that sphere is disturbed. Even though such a disturbance may be small, it cannot in general be neglected—in fact, the inverse square law conspires against such negligibility, for a rather interesting reason (see appendix 7.3).

This sensitivity to small changes—to differences in "initial conditions"—provides one way for a time-irreversible universe to obey time-reversible laws. Initial conditions that are time-asymmetric break the time-reversal symmetry, and the entire dynamical trajectory can then be time-asymmetric too. Figure 7.1 is a schematic representation of the phase space of an N-particle system. Here $q = (q_1, \ldots q_N)$ lists the position coordinates of the particles, and $p = (p_1, \ldots p_N)$ lists the velocity coordinates. The dimensions of q and p have ben reduced to 1 for ease of drawing. The time-reversal operator sends (q,p) to $(q,-p)$; that is, it simultaneously reverses all velocities while leaving positions unchanged. In the picture this operator can be visualized as a top/bottom reflection, or "flip." The q-axis (given by $p = 0$) is fixed by this flip.

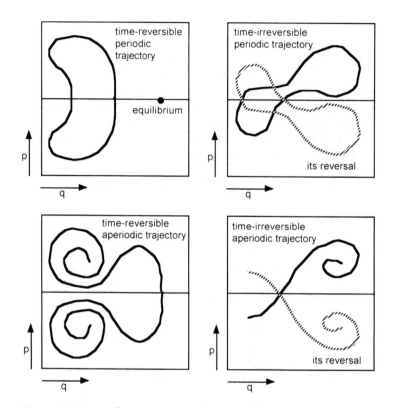

Figure 7.1 Types of trajectories in a time-reversible system.

Equilibrium states have $p = 0$ and so are time-reversible—the reverse of "do nothing" is "do nothing." Periodic cycles can be time-reversible, in which case the entire trajectory is transformed into itself by the flip. Even so, the direction of the "arrow of time" along the trajectory is reversed. Periodic cycles can also be time-irreversible; the flipped cycle is different from the original. Aperiodic trajectories can be time-reversible, in which case the entire trajectory is again transformed into itself by the flip, but the direction of the "arrow of time" along the trajectory is reversed. Aperiodic trajectories can also be time-irreversible. Note that any trajectory along which entropy changes monotonically cannot be periodic, unless entropy is constant along that trajectory. Moreover, any trajectory that hits the q-axis must be a single point, so the apparent intersections of trajectories with the q-axis are artefacts of the dimension reduction.

This property of initial conditions explains why a time-asymmetric universe *can* obey time-symmetric laws, but it doesn't explain why the time-asymmetry of our universe always seems to "go in the same direction." (Does it? Always? Not so clear—see Cohen & Stewart 1994.) In particular, according to the second law of thermodynamics, *all* states of a gas tend to equilibrium along the direction of entropy-increase. Even if we admit Boltzmann's point that this law is valid only statistically, there is still the usual difficulty. For every state of a gas that follows the direction of entropy-increase (type I) there is a dual one that follows the direction of entropy-decrease (type D). Figure 7.2 again shows the phase space of an N-particle system. Initial conditions have been divided into two types: ordered and disordered. Assume that if a state (q,p) is ordered, then so is its flip $(q,-p)$. There are two reasons for making this assumption. The first is that it is natural; simultaneously reversing all velocities should not make an ordered state become disordered. (Indeed, order is typically described purely in terms of *configuration q*.) The other is that if this assumption is invalid, then the definition of "order" has time-asymmetry built into it, so it is not surprising if it leads to time-asymmetric conclusions.

Assume that to every state we can associate a number, its entropy. By the preceding argument, the entropy of a state and of its flip should be equal. If we define a measure by dividing (q,p)-space into equal-

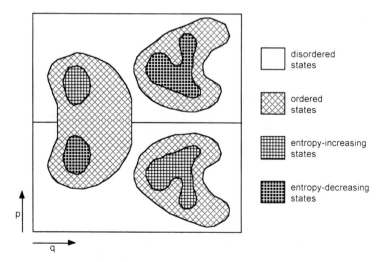

disordered states

ordered states

entropy-increasing states

entropy-decreasing states

Figure 7.2 Types of state in an N-particle system.

sized boxes, then the measure of a set is the same as that of its flip. Indeed, any sensible measure that we place on the phase space, to determine the probability of encountering a given state, should be preserved by the flip—otherwise we have (literally) loaded the dice in favor of time-asymmetry of entropy change.

Among the ordered states, *some* have trajectories with increasing entropy. These cannot be *all* ordered states, because their images under the flip automatically have trajectories with decreasing entropy. Unless entropy is constant on all trajectories originating from ordered states— in which case again there is no problem to discuss—we find that type I states and type D states are paired by the flip. All of this works even if there are type C states or entropically more complex ones. So, as the figure suggests, type I and D states are "mirror images" of each other under time-reversal. Therefore, among those states whose entropy-trajectory is time-asymmetric and monotone, half the states of a gas follow the direction of entropy-increase, while the other half follow the direction of entropy-decrease. So initial conditions won't do the trick— unless some new physical principle selects the initial conditions preferentially for one direction of entropy-flow.

No, but Coarse-Graining Is

To explain why thermodynamics works so well as a description of many phenomena in our universe, we must dig deeper. Is there some kind of bias in the choice of initial states, for example, that makes the entropy-increasing ones much more "natural" as a description of our universe? Possibly. You can make out quite a plausible case: The states that can be described compactly, those with *localized causes*, like "Drop a watch on the floor," are generally of the entropy-increasing kind. In contrast, entropy-decreasing states require elaborate descriptions—"Take all the components of a broken watch, propel each one in some direction at some particular speed, set up a certain complex pattern of vibrations in the floor . . . do all this just right," and you can reassemble the watch. It might be worth thinking about this idea more deeply, since "compact description" lies at the basis of Chaitin's algorithmic information theory.

Here, however, I want to emphasize a different source of time-asymmetry: modeling assumptions. The analysis of both systems— thermodynamic and gravitic—involves an assumption of *coarse-graining* (and subsequent smoothing, but that is more harmless). The coarse-graining assumption is "realistic" for both systems, which is why they are both good physics; but *it is realistic in different domains of discourse.* This means that we are not free to ignore the modeling assumptions and describe events as if these assumptions are simply "true." In particular, we cannot make both assumptions in the same discourse and expect to get anything sensible out of it, which is what Penrose attempts to do.

Coarse-graining introduces a *modification* of the original time-reversible setup, because fine-scale differences of states are ignored. This is right and proper for the intended purpose—steam engines and their ilk—but it is also responsible for time-asymmetry. Let me unpack this asymmetry. Two states that are distinguishably different *despite* coarse-graining can, in some circumstances, evolve over the passage of forward time into states that are indistinguishable in the coarse-grained approximation. (Proof: Forget the coarse-graining and work with the infinite-precision model. Take two states that are equivalent up to coarse-graining; evolve them backward in time until their differences reveal themselves at the coarse-grained level; now take those as your two states, put back the coarse-graining, and watch them evolve to *identical* states in the coarse-grained world. This proof would not work if the dynamic flow preserves the coarse-graining, but it does not.)

Notice that in the coarse-grained model this extra assumption—that differences below the coarse-grained level become indistinguishable *and remain indistinguishable for all forward time*—is explicitly time-asymmetric. New differences cannot "bubble up from" the coarse-graining (as they did in the preceding reverse-time proof) if we remain in the coarse-grained world (as the proof did not).

This is one way that mathematicians make respectable the approximations inherent in the traditional thermodynamic approach to entropy. It does not explain why entropy increases in closed thermodynamic systems—that requires further analysis, the usual stuff in the thermodynamics textbooks—but it does explain why a time-asymmetric conclusion of that type is possible. Coarse-graining destroys time symmetry of the model. It picks *one* arrow of time, the thermodynamic one. The importance of coarse-graining in this context was known to Boltzmann (1909), Gibbs (1902), and Jauch and Báron (1972), yet many authors ignore its implications.

The main implication is that the breaking of time symmetry need not be a consequence of initial conditions. In fact, in thermodynamics, temporal symmetry-breaking has a very different origin. We have already seen that at the microlevel, trajectories of types I and D are temporal "mirror images" of each other; unless we wish to load time's dice, then whatever measure we assign to the initial conditions for type I, we must assign the same measure to type D. The situation with coarse-graining is quite different. Here there are a set of initial macrostates of measure very close to zero for which entropy fails to increase and a complementary set of macrostates (*all* the others, measure 1) for which entropy increases. There is *no* mirror-image set of macrostates for which entropy decreases. This situation is very different from the one proposed by Penrose.

In short, there are at least three ways to introduce asymmetry into a time-reversible theory.

1. By restricting to a time-asymmetric class of interactions
2. By choice of initial conditions
3. By coarse-graining phase space

Method 1 removes any element of paradox. Method 2 fails because of the mirror-image duality between forward and backward time at the microstate level. This leaves method 3. Physics, which, unlike mathematics, has the option of changing from one theory to another during the same analysis of reality, employs all three, generally tacitly. The potential for confusion is clear.

This is by no means the end of the story. To me the biggest puzzle is that we live in a universe where it seems easy to set up initial conditions for entropy-increasing interactions (drop a cup and watch it smash) but hard to set up the reversed initial conditions (take a broken cup and propel its parts together in such a way that they rejoin). This makes the modeling assumptions of thermodynamics widely useful as a description of reality. Why is the universe like this? Is it? (What about a lump of rock aggregating on the bed of a river? What about a crystal growing in a solution?) Part of the explanation is that the initial conditions that can most readily be set up in this universe are those with localized causes, whereas time-reversed situations require initial conditions that involve global coordination of spatially separated events. This topic could be explored at length; it is where the heart of the problem of time-reversibility *really* lives. However, I must now move on to gravitation.

Gravitic Clumping and Symmetry-Breaking

The physicists' description of the changing state of the universe as it clumps under the force of gravity also involves coarse-graining. The source of this tacit coarse-graining is the statement that the early universe is "smooth." However, the early universe consists of radiation, (and later of particles, where the coarse-graining is even clearer), and the quantum wave-function of a photon (say) is *not* spatially uniform, hence *not* smooth on very tiny scales. To say that the universe is smooth amounts either to coarse-graining it and saying that there is no discernible difference between the grains or (much the same thing) taking a statistical stance.

What makes gravitation so interesting—and is generally accepted as the cause of clumping—is that "uniform" systems of gravitating bodies are *unstable* (appendix 7.3). This means, quite specifically, that differences smaller than any specific level of coarse-graining not only *can* "bubble up" into macroscopic differences but almost always *do*. This is the big difference between gravitics and thermodynamics. The thermodynamic model that best fits our universe (for very long but perhaps not infinite periods of time—it models "long-term transients," which are what we actually observe) is one in which differences can dissipate by disappearing below the level of coarse-graining as time marches forward. The gravitic model that best fits our universe (in this case perhaps even for infinite time—not transients) is one in which differences can amplify by bubbling up from below the level of coarse-graining as time marches forward. The relation of these two scientific domains to coarse-graining is exactly *opposite* if the same arrow of time is used for both.

I can now explain the "entropy gap" between the early and late universes, observed by Penrose and credited by him to astonishingly unlikely initial conditions. It has nothing to do with initial conditions; *it is an artefact of coarse-graining*. Gravitational clumping bubbles up from a level of coarse-graining to which thermodynamic entropy is, by definition, insensitive. If you insist on thinking in thermodynamic terms, you could say that gravitation by its nature creates negative entropy in forward time, but my feeling is that the coarse-graining of thermodynamics makes it a rather blunt instrument for discussing such questions. Nor is there any reason for believing that the transition from gravitics to negative entropy can be quantified, because it is crucially dependent on structure below the level of the coarse-graining. It may be fruitful to view the universe as a "complicity" between thermodynamics and gravitics in the sense of Cohen and Stewart (1994), in which case all of the really interesting phenomena are emergent—even if they are consequences of some extended system of laws, there will be no practical way to derive those consequences algorithmically. This may not appeal to physicists, but it will be rather attractive to biologists, who find physical laws somewhat unilluminating when it comes to studying living organisms.

There is an attractive "information" interpretation of everything I've been saying, and I'll mention it even though I believe that most discussions of information in physics suffer from the same inconsistencies as in the use of coarse-graining. The idea is to see coarse-graining as a "loss of information" about states in phase space. Coarse-grained states are fuzzy, less precisely defined—their coordinates fuzz out after finitely many bits, whereas a finer level of description brings more bits into play. So we can view the increase of entropy (in a coarse-grained model) as a flow of information from coarse scales to fine ones—this is where the missing "negentropy" *goes*. Similarly, the gravitational increase of clumpiness arises because gravity pulls information out of the small scales and makes it visible on larger scales. I think these statements are at best puns (see appendix 7.1), but they may possess mathematical validity in some suitably limited realm of discourse.

Both statements, even if true, are statistical. The increase of entropy in classical thermodynamics, as is widely recognized, is a statistical law. All the molecules of air in a box *could* accumulate in one half; however, this is extremely unlikely. The "dual" negentropic nature of gravity is also statistical, for similar reasons. There are clear physical differences, however: elastic collisions are short-range and repulsive, whereas gravity is long-range and attractive. With such different force laws, it is hardly surprising that the generic behavior should be so different. (As an extreme case, imagine systems where "gravity" is so short-range that it has no effect unless particles collide, but then they stick together forever. Increasing clumpiness would then be obvious.)

The real universe is both gravitational and thermodynamic. In some contexts, the thermodynamic model is more appropriate and thermodynamics provides a good model. In other contexts, a gravitational model is more appropriate. There are yet other contexts; molecular chemistry involves different types of forces again. It is, I think, a mistake to attempt to shoehorn all natural phenomena into the thermodynamic approximation *or* into the gravitational approximation. Both *are* approximations, based on coarse-graining. It is especially dubious to expect both thermodynamic and gravitic approximations to work in the *same* context, when the way they respond to coarse-graining is diametrically opposite.

A Fourth Law for Life?

Can we stretch the gravitic insight far enough to get a biological one? I think so. Let me advance a rough analogy. The biosphere can be thought of as a jazzed-up kind of gravitic system, in which "organisms" or "autonomous agents" play the role of masses and their interactions play the role of the law of gravity. In order for the coarse-graining assumption behind thermodynamics to apply to such a system, a "uniform" or "equilibrium" state must be *stable*. For if it is unstable, then structure can—indeed typically *will*—"bubble up" from below the level of coarse-graining, whereas thermodynamics assumes this does not happen.

Now, it is in the nature of the biosphere (and also a consequence of Kauffman's notion of "autonomous agent") that a (statistically) equilibrium state should be unstable. A successful predator, for example, will deplete the prey in its vicinity. Evolution will subvert any obvious patterns, and there are few patterns more obvious than uniformities. Indeed, the entire rich panoply of life on Earth argues for the instability of the uniform state. This instability leads to symmetry-breaking of the uniform state, hence to the development of "order" out of disorder. This "order" grows, locally in space and time. Globally, it may continue to grow, or it may disappear. However, even if the system returns after some time to a uniform state, structure will again bubble into existence.

This kind of emergence of "order" is therefore unavoidable. It is *not* the result of absence of closure (the universe as a whole is closed). It is *not* the result of special initial conditions (on the contrary, it is typical when autonomous agents exist). In particular it is *not* a statistical freak, occurring with negligible probability. It is not consistent with the second law of thermodynamics, but this is *not* because it somehow "breaks" the law. It is completely outside the law's jurisdiction: when structure can bubble up from below any specific level of coarse-graining, the second law simply *does not apply*. If a law doesn't apply, you can't break it.

A Second Law of Gravitics?

To summarize. The "laws" of thermodynamics, especially the celebrated second law, are statistically valid models of nature in a particular set of contexts. They are *not* universally valid truths about the universe (as the clumping of gravity demonstrates). Given the strong "time-reversal duality" between coarse-graining in thermodynamics and coarse-graining in gravitation, it seems plausible that a gravitational complexity measure analogous to entropy in thermodynamics might one day be constructed, along with a "second law of gravitics" to the effect that the complexity of a self-gravitating system of particles increases with time. (The most obvious technical obstacle is the long-range nature of gravity.) This will also be a statistically valid model of nature in a particular set of contexts. I find it interesting that even though coarse-graining works in opposite ways for these two types of system, both "second laws" would correspond rather well to our own universe. The reason is that I have biased the formulation of the laws in terms of what we observe in our own universe as time runs in its normal direction. Nevertheless, despite this apparent concurrence, the two laws would apply to drastically different physical systems: one to gases, the other to self-gravitating systems.

The complexification of self-gravitating systems is currently the best candidate for a "law of increasing complexity." If there is any truth in what I've just discussed, the conclusion will have to be that any "laws of complexity" will *also* have to be statistically valid models of nature in a particular set of contexts. Different physical systems would fit some contexts but not others; some of the laws would be valid and others invalid, depending on the system.

This is not as "clean" a structure as many complexity theorists hope for. Would it count as a "fourth law of thermodynamics"? It isn't really in the thermodynamic frame at all. On the other hand, there is no good reason to confine ourselves to the thermodynamic frame in any case, for if we do, we can't understand gravitic clumping, let alone organisms self-complicating. Possibly it is the cleanest we will get: the "glass menagerie" of Cohen and Stewart (1994) in which nature is described by a collection of different rule-systems with limited, sometimes overlapping, domains of validity ("jurisdictions").

For those who like grand syntheses, though, there is another, far more elegant, possibility. Perhaps all those different "second laws" are all special cases of a single, universal second law—but a law that is *context dependent*. Given a physical system and a set of forces acting upon it, we plug both the system *and* its context (the appropriate set of forces, maybe more; its "jurisdiction") into the universal law and read off what will happen. Maybe there does exist a fourth law of thermodynamics in this sense—a generalization, not an extension. After all, the main difference between the second law of thermodynamics and my hypothetical second law of gravitics is just a change of sign.

Summary

The tendency of a thermodynamic system to become increasingly disordered has been contrasted with that of a gravitic system to become ever more clumpy as time passes. The physical difference between the two is that a thermodynamic system can be modeled as a system of particles acted upon by short-range repelling forces, whereas a gravitic system can be modeled as a system of particles acted upon by long-range attracting forces. This leads to very different behavior, and the assumption that both models are simultaneously "true" mostly leads to nonsense, since both are approximations to reality and their domains of validity differ considerably.

The second law of thermodynamics says that the entropy of a system—usually interpreted as its "disorder"—almost always increases as time passes. However, the underlying classical dynamics of the molecules in a gas is time-reversible: If some sequence of events is consistent with the laws of motion, so is the time-reversal of that sequence. The relation between these two physical principles appears paradoxical, and it can be explained in a number of ways. Usually it is explained away by appealing to the statistical nature of the second law of thermodynamics or by a special choice of initial conditions that selects a particular arrow of time. However, neither of these ideas adequately explains the observed behavior of our current universe. The real source of the paradox is a "coarse-graining" step involved in the deduction of the second law of thermodynamics, in which motion below a certain scale of measurement is ignored. Although this ignorance *can* be interpreted as "missing information," such an interpretation is fraught with difficulties and is probably best avoided. All teminology in this area, including "order," "disorder," and "information," is loaded. Most terms have several different interpretations, and many commentators slide from one interpretation to another without exercising due care.

The differences between these two "toy" systems illuminate the quest for a fourth law of thermodynamics, able to formalize or explain the tendency of living systems to increase their degree of order. In particular these differences suggest that any such law must be formulated in a context-dependent manner and not simply adjoined to the first three laws of thermodynamics on the assumption that such laws are universally valid. Kauffman's proposals about autonomous agents are suitably context dependent.

The interactions in living systems can be long- or short-range, and repulsive or attractive. They are also heavily context dependent. Naïve extrapolations from toy systems that have simple interactions and lack richness of context are likely to be misleading.

Appendix 7.1. Negentropy and Disinformation

In information theory (Shannon & Weaver 1964), there is a quantity called *entropy* that is the negative of information. Moreover, the formula for entropy is essentially the same as Boltzmann's formula for thermodynamic entropy in terms of numbers of microstates compatible with a given macrostate (Planck 1945; Sommerfeld 1964). It is therefore tempting to think of thermodynamic entropy as negative information, and many authors do just that. Indeed, the increase in thermodynamic entropy of a system can be viewed as a loss of information: details of the motion disappear below the level of the coarse-graining and are thus "lost."

Whether it is actually meaningful to view thermodynamic entropy as negative information has been a controversial issue for many years. It is related to some interpretations of "Maxwell's demon," nowadays generally conceived as a superintelligent being that can sense the motions of individual molecules in a gas without affecting them and then react to them (see Leff & Rex 1990). Such a demon can open or close a trapdoor in a dividing wall, letting only fast-moving molecules through: the gas on one side of the wall then heats up while that on the other side cools, contrary to the second law of thermodynamics. Szilard (1964) pointed out that in practice the necessary measurements generate entropy, which neatly cancels out the reduction in entropy accomplished by the demon. Brillouin (1949) considered what happens if the demon uses light signals to gain information about the molecular motions. The light radiation is generated by some external source of energy, which takes the system out of thermodynamic equilibrium and "pours negative entropy into the system." Thus to Brillouin, information means negative entropy. Raymond (1951) gave a more detailed argument in support of the same view. Ter Haar (1995) warns against interpreting information as a thermodynamic quantity, while pointing out that entropy can be associated with loss of information in a loose sense. Jauch and Báron (1972) quote ter Haar and argue firmly against any identification of information with negative thermodynamic entropy. DeBeauregard and Tribus (1974) dispute Jauch and Báron's conclusions. The disagreements rumble on, even today . . .

What seems to be happening here is a confusion between a formal mathematical setting in which "laws" of information and entropy can be stated, a series of physical intuitions about heuristic interpretations of those concepts, and a failure to understand the role of context. Much is made of the resemblance between the *formulas* for entropy in information theory and thermodynamics; little attention is paid to the context to which those formulas apply. One important difference is that in thermodynamics, entropy is a function defined on *states*, whereas in information theory it is defined for an information *source*—a system that generates ensembles of states ("messages"). Roughly speaking, this is the phase space for successive bits of a message, and a message is a trajectory in phase space. A specific configuration of gas molecules has a thermodynamic entropy, but a specific message does not. This fact alone should serve as a warning. Even in information theory, the information "in" a message is not negative information-theoretic entropy. Indeed, the entropy of the source remains unchanged, no matter how many messages it generates. The situation is not quite as clear-cut as these comments suggest, however: it is possible to "transfer" the information-theoretic entropy measure from the source to individual messages by assuming that the messages are generated by some specific stochastic process (a Markov chain is commonly assumed). Rather than analyze this trick in detail, I shall just observe that it is highly context dependent.

Those authors who argue that information is negative thermodynamic entropy do so in some specific, carefully chosen context. If one unravels their argument, it usually turns out that their chosen measure of information is tailored to transfer thermodynamic entropy into the system from some external source. This transfer permits the selection of states in terms of structure that is below the level of coarse-graining (which is how all "Maxwell's demons" work) and can therefore be viewed as "obtaining extra information." The authors who then argue against the resulting findings represent this source by some thermodynamically reasonable system—that is, by a genuine mechanism such as a torch shining a beam of light or a mechanism defined on the scale where coarse-graining is legitimate (which is, in particular, the scale appropriate to all human-made machines, but maybe not to biological ones, which affect individual molecules—for example, pores in membranes, suich as ion channels). They then show, often by detailed calculations, that if this source and the original system are combined, the resulting system is thermodynamically closed, and there is then no reduction in thermodynamic entropy. There is nothing objectionable about both sides of the argument as long as the nature of the game is appreciated; on the other hand, the value of repeatedly playing this game in specific detail when the general principle behind it is so transparent must be questioned.

Appendix 7.2. Order and Disorder

The issues addressed in this chapter are commonly discussed in terms of "order" and "disorder," as if these concepts have a unique and well-defined meaning that agrees with naïve intuition in all respects. (For example, it is taken for granted that a state of a gas with all molecules in one half of a box is "ordered," whereas a state in which the box is uniformly filled is "disordered." However, it is easy to give counterexamples with "implicate order," such as a state that looks "disordered" *now* but will become "ordered" in one hour's time.) Thermodynamic entropy is metaphorically identified with disorder. However, the assumptions involved in such terminology are exceedingly dubious.

To appreciate just how dubious, consider what is perhaps the simplest known chaotic dynamical system, the *quadratic map* (or *logistic map*)

$$x_{t+1} = kx_t(1 - x_t)$$

where $0 \leq x \leq 1$ and k is a constant between 0 and 4. This is a discrete-time dynamical system, and any initial value x_0 generates a trajectory $x_1, x_2, \ldots$ by iterating the formula. The dynamic behavior depends on k (see May 1976). When $k < 3$ the system settles to a steady state. For $3 < k < 3.5$ it settles to a period-2 point. For larger k there is a "period-doubling" cascade of states of period 4, 8, 16, 32, ... culminating in a chaotic state for $k \sim 3.57$. Broadly speaking, as k increases, the system becomes ever more chaotic—though, as we will soon see, there are exceptions.

Mathematicians call variables like k *bifurcation parameters*: Changing their values may result in changes to the dynamic behaviour (bifurcations). Physicists, especially those working in nonequilibrium thermodynamics and synergetics, call variables like k *order parameters*: Changing their values may result in changes from order to disorder. This terminology is loaded, as becomes apparent if we plot the *bifurcation diagram* of the logistic map (fig. 7.3), showing how the long-term dynamics ("attractor") varies with k. We observe that in the chaotic regime $k >$ 3.57 there exist innumerable "windows" of nonchaotic, indeed periodic, behavior. For example, a period-3 window sets in around k = 3.835, and a period-5 window around k = 3.739. Mathematically, it can be proved that infinitely many such periodic windows exist.

There is more. Within these periodic windows are subwindows of chaos. Indeed, within the period-3 window we can find small copies of the entire original bifurcation diagram. The interpretation of the parameter k as something whose increase leads from "order" to "disorder"' makes very little sense, for regimes of order and disorder are fractally intertwined, in an infinite cascade. While there may be nothing *technically* incorrect in adopting the order/disorder terminology here, it seems clear that the original intuition that led to the adoption of this terminology comes to pieces when confronted with the simplest ex-

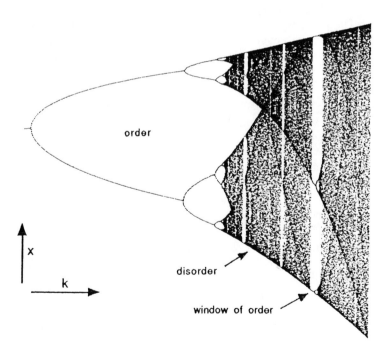

Figure 7.3 How the dynamical attractors of the quadratic map vary with the parameter k. Curves represent regions of "order"; irregularly dotted regions represent "disorder."

ample of a transition to chaos. Indeed, since periodic and chaotic behavior in this example are so intimately entwined, it seems naïve to equate chaos with disorder and periodicity with order. In terms of the geometry of the quadratic map, they seem pretty much on the same footing, and the distinction between them seems to be an artefact of paying too much attention to *things* when what is going on is *processes*. Again, this points toward Chaitin's algorithmic information theory.

Appendix 3. A Qualitative Second Law of Gravitics

The "law" is:

> *As time passes (forwards according to the thermodynamic arrow), gravitic systems tend to become more and more clumpy.*

Observations support this law—but what causes the clumping?

The inverse square law of Newtonian gravitation is beautifully "fine-tuned" to cause a uniform distribution of matter in three–dimensional space to be unstable. Specifically, perturbations of the uniform state that cause it to clump do not die away but create further clumping. As in this entire subject, there are substantial technical obstacles if we wish to make such statements rigorous (see Binney & Tremaine 1987, chaps. 4 and 5), but the essential qualitative ideas are relatively simple. Suppose that space is filled with an approximately (that is, statistically) uniform distribution of gravitating bodies. Choose a specific body S and consider the effect on S of bodies that lie within some double-cone with S as vertex—which intuitively represent the bodies that attract S in roughly the direction of the axis of the cone (with either orientation).

Consider two "clusters" of bodies, defined to be the intersections V_1, V_2 of this cone with two spherical shells of bodies of equal thickness T, which we assume to be small. One shell is at distance r_1 and the other at a larger distance r_2 from S (fig. 7.4). We ask how the gravitational attractions of these two clusters compare. The volume of V_1 is approximately cTr_1^2, for a constant c, and the volume of V_2 is approximately cTr_2^2, for the same constant c. (This constant represents the "solid angle" at the vertex: the proportion of any spherical surface centered on S that lies within the cone.) Assuming that the density of bodies is uniform, the masses of the bodies in V_1, V_2 are proportional to the volumes, hence equal to ar_1^2, ar_2^2, for some constant a. The attractive forces these clusters of bodies exert on S are, respectively, ar_1^2/r_1^2,

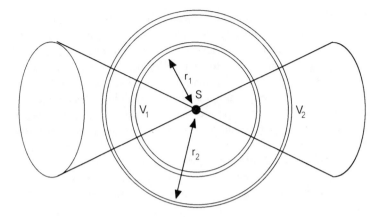

Figure 7.4 The gravitational forces exerted on a body S by equally dense regions at different distances, but subtending the same solid angle, are equal.

ar_2^2/r_2^2, which both equal a. In other words, the more distant cluster exerts the same force on S as the nearer one—but in this case in the opposite direction. Thus the forces exerted on S by V_1 and V_2 cancel out.

Finally, suppose that the density of bodies in V_2 is greater than that in V_1, so that V_2 forms a denser cluster. Then the attraction of V_2 is greater than that of V_1, so there is a net attraction on S in the direction of the denser cluster. Thus (assuming all forces act instantaneously) the acceleration of a given body is determined by large-scale gradients in the density of bodies within the entire universe but *not* by the local density. This phenomenon occurs because Newtonian gravitation *decreases* as the square of the distance, while the volume of a spherical shell of bodies of fixed thickness *increases* as the square of the distance. The two neatly balance, so that in a uniform distribution of bodies the attractive effect of all bodies within a given angular region is independent of distance. In consequence, *a uniform distribution is unstable*; regions of higher density than the average exert a net attraction on all other bodies, so nascent clusters tend to grow.

The situation is still quite delicate, however. Again assume a uniform distribution of bodies. By essentially the same argument, if some (spherical) region of bodies collapses down to a very small "point" mass, it exerts the same force on all other bodies as if it had not collapsed at all. This fact was first discovered by Isaac Newton, as the principle that a spherical planet exerts the same gravitational force (on its exterior) as an equal mass concentrated into a point at its center. Therefore, *a distribution that is statistically uniform at some length scale is* also *unstable*. By "statistically uniform at some length scale" I mean that any region of space whose diameter is equal to the length scale contains approximately the same total mass. Suppose, then, that we have a uniform distribution of bodies and we perturb it to get many random clusters, statistically uniformly distributed on larger spatial scales. Let the clusters collapse under gravity (mathematically it helps to assume a small amount of friction to dissipate some of their energy). Then we end up with a statistically uniform distribution of point masses again. Therefore, the same clumping process will happen on the larger spatial scale, provided there exist perturbations on that scale. That is, the law of gravity has a *scaling symmetry*: if distances multiply by r and masses by m, then all forces multiply by m/r^2. Starting from a uniform distribution of matter, we are led to expect a fractal hierarchy of clumping events on larger and larger scales. This kind of behavior has been called *oligarchic growth* when it occurs on the scale of solar systems (Kokubo & Ida 1990), but since the law of gravity has a scaling symmetry we may employ the concept on any scale. Thommes, Duncan, and Levison describe the formation of solar systems in these terms:

> The most plausible model for the formation of giant-planet cores in the Jupiter-Saturn region is based on the concept of oligarchic growth. In this model, the largest few objects at any given time are of comparable mass, and are separated by amounts determined by their masses and distances from the Sun. As the system evolves, the mass of the system is concentrated into an ever-decreasing number of bodies of increasing masses and separations. (1999, 635)

Exactly.

REFERENCES

Auyang, S.Y. 1998. *Foundations of Complex-System Theories*. Cambridge: Cambridge University Press.

Baez, J. C. 1994. Review of *The Physical Basis of the Direction of Time*. by H. D. Zeh. *Mathematical Intelligencer* 16,1: 72–75.

deBeauregard, O. C., and M. Tribus. 1974. "Information Theory and Thermodynamics." *Helvetica Physica Acta* 47: 238–247.

Bennett, C. H. 1973. "Logical Reversibility of Computation." *IBM Journal of Research and Development* 17: 525–532.

Bennett, C. H. 1982. "The Thermodynamics of Computation—A Review." *International Journal of Theoretical Physics* 21: 905–940.

Binney, J., and S. Tremaine. 1987. *Galactic Dynamics*. Princeton: Princeton University Press.

Bohm, D. 1980. *Wholeness and the Implicate Order*. London: Routledge and Kegan Paul.

Boltzmann, L. 1909. *Wissenschaftliche Abhandlungen von Ludwig Boltzmann*. Edited by F. Hasenorhl. Leipzig.

Brillouin, L. 1949. "Life, Thermodynamics, and Cybernetics." *American Scientist* 37: 554–568.

Chaitin, G. 1987. *Information, Randomness and Incompleteness*. Singapore: World Scientific.

Cohen, J., and I. Stewart. 1994. *The Collapse of Chaos*. New York: Viking Press.

Daub, E. E. 1970. "Maxwell's Demon." *Studies in the History and Philosophy of Science* 1: 213–227.

Davies, P. 1992. *The Mind of God*. New York: Simon and Schuster.

Dawkins, R. 1987. *The Blind Watchmaker*. New York: Norton.

Dyson, F. 1979. "Time without End: Physics and Biology in an Open Universe." *Reviews of Modern Physics* 51: 447–460.

Gibbs, J. W. 1902. *Elementary Principles in Statistical Mechanics*. New Haven: Yale University Press.

ter Haar, D. 1995. *Elements of Statistical Mechanics*. Oxford: Butterworth-Heinemann.

Hawking, S. 1988. *A Brief History of Time*. New York: Bantam.

Jauch, J. M., and J. G. Barón. 1972. "Entropy, information, and Szilard's Paradox." *Helvetica Physica Acta* 45: 220–232.

Kauffman, S. A. 1993. *The Origins of Order*. New York: Oxford University Press.

Kauffman, S. A. 1995. *At Home in the Universe*. New York: Oxford University Press.

Kauffman, S. A. 2000. *Investigations: Scope for a Possible "Fourth Law" of Thermodynamics for Non-Equilibrium Systems*. New York: Oxford University Press.

Kokubo, E., and S. Ida. 1990. "Oligarchic Growth of Protoplanets." *Icarus* 127: 171–178.

Landauer, R. 1961. "Irreversibility and Heat Generation in the Computing Process." *IBM Journal of Research and Development* 5: 183–191.

Leff, H. S., and A. F. Rex, eds. 1990. *Maxwell's Demon: Entropy, Information, Computing*. Bristol, England: Adam Hilger.

Loschmidt, J. 1869. "Der zweite Satz der mechanischen Wärmetheorie." *Akademie der Wissenschaften, Wien, Mathematisch-Naturwissenschaftliche Klasse, Sitzungsberichte* 59, 2: 395–418.

May, R. M. 1976. "Simple Mathematical Models with Very Complicated Dynamics." *Nature* 261: 459–467.

Planck, M. 1945. *Treatise on Thermodynamics*. New York: Dover.

Powell, C. S. 1994. "Sanity Check," *Scientific American*, June: 9–10.

Raymond, R. C. 1951. "The Well-informed Heat Engine." *American Journal of Physics* 19: 109–112.

Shannon, C. E., and W. Weaver. 1964. *The Mathematical Theory of Communication*. Urbana: University of Illinois Press.

Sommerfeld, A. 1964. *Thermodynamics and Statistical Mechanics*. New York: Academic Press.

Stewart, I. N., and M. Golubitsky. 1992. *Fearful Symmetry—Is God a Geometer?* Oxford: Blackwell.

Szilard, L. 1964. "On the Decrease of Entropy in a Thermodynamic System by the Intervention of Intelligent Beings." *Behavioral Science* 9: 301–310.

Thommes, E. W., M. J. Duncan, and H. F. Levison. 1999. "The Formation of Uranus and Neptune in the Jupiter-Saturn Region of the Solar System." *Nature* 402: 635–638.

Toffoli, T. 1982. "Physics and Computation." *International Journal of Theoretical Physics* 21: 165–175.

Zeh, H. D. 1992. *The Physical Basis of the Direction of Time*. New York: Springer.

EIGHT

TWO ARROWS FROM A MIGHTY BOW

Werner R. Loewenstein

Our Awareness of Time

Aristotle's maxim "Nothing is in the mind that did not pass through the senses" may be a little strong for today's tastes, but there is more than a grain of truth in it. We see, we hear, we feel, we are aware, at least in large part, thanks to our senses. But what precisely do we mean when we say we are aware of something? What is this peculiar state we call consciousness? In recent years information theory has given us fresh insights into the question of cognition, and neurobiology has offered some views of the underlying sensory information stream. But the highest levels of that stream, the information states leading to consciousness, have not been penetrated. This is frontier, perhaps the last true frontier of biology—a natural meeting-ground with physics.

I will address myself to one aspect of that mystery, our sense of time. This sense is a prominent feature of consciousness, and I mean here not the sensing of periodicities and rhythms (the causes of which are reasonably well understood) but something more fundamental: the sensing of time itself, the passing of time. This we feel as something intensely real. We feel it as a constant streaming, as if there were an arrow inside us pointing from the past to the future.

This arrow is an integral part of our conscious experience and, more than that, a defining part of our inner selves. Yet, for all its intimacy and universality, our sense of time has defied scientific explanation. Not that this is the only perverse thing regarding consciousness, but I single out this one because our sense of time appears to be the least ethereal and, with some prodding, it might give ground. It has a measurable counterpart in physics, the quantity t. Indeed, that t occupies a high place in our descriptions of the physical universe—it enters in all equations dealing with events that evolve. However, we should be forewarned that that time and the one we sense may not be the same; t is a sort of housebroken variety of time, a variety that got tamed through mathematics. Nevertheless, it still has some of its wild qualities and, for something that for hundreds of years has gone so thoroughly through the wringer of mathematics, it is surprisingly human.

But what is this thing, at once so utterly familiar and so bewilderingly mysterious, we call time? Saint Augustine, one of the early inquirers and among the most profound, captured the zeitgeist when he wrote at the turn of the fifth century in his *Confessions* (book 6.14.17), "If no one asks me, I know; but if someone wants me to explain it, I don't know" (Augustine 1991, 230). He might as well have said this today. We are still groping, and if we ponder the question in terms of current physics, we may well wonder why time should have an arrow at all.

The roots of our sense of time are somehow interwoven with those of our conscious states and can be traced to the transience of these states—their constant flitting, one state following another. Ordinarily, those states are so many and follow each other so fast that one doesn't see the forest for the trees. But that gets better if one fixes on a particular sequence—one streamlet in the stream. Try this, for example: have someone tap the tip of your index finger rhythmically with a pencil, and have him space the taps half a second apart. You will feel a series of touch sensations that jointly convey a forward sense of time. The nerves in your finger send the tactile information to the brain; it takes somewhat under 50 milliseconds for that information to get to the cortex and some 200–300 milliseconds for it to be processed and become a conscious state—enough time for an unslurred sequence of events at the various stages. Now, change the condition slightly and tag the sequence by making the second tap stronger than the first. You will then feel the sensations in a distinct order, an outcome as revealing as it is plain: The conscious states are well ordered, and the order reflects that of the peripheral stimuli.

The same can be said for the states underlying our auditory and visual experiences. A series of notes played on the piano are heard in the sequence they were played, and a series of pictures flashed on a TV screen are seen in the same serial sequence. There are limits to the speed with which our nervous system can handle and resolve individual events; but *within these limits, our conscious states always occur in unjumbled sequence.*

If those experiments are too homespun for your taste, ask a neuroscientist to upgrade them by tapping the pertinent signal flow on its way to the brain or inside the brain—today's kibitzers can listen in digital splendor. But regardless of how the experiments are done, the conclusion is the same: The order in the sensory stimuli is preserved in the conscious states.

This orderly sequencing prevails also in conditions of multiple sensory input. We see the finger movements of a pianist in the same sequence as the notes they produce, or we hear the taps of the tap dancer in step with her movements. Not even an illusion—such as hearing an extra note when there was none or seeing a phantom dance step—will jumble the picture. The bogus information is merely inserted into the sequence. And higher up, when memory comes into play, the imagined information reels off with the true (the bane of our courtrooms).

I bring up the matter of bogus images here because it bears on our intrinsic thought processes, by which I mean the conscious states originating from within our brain, in the absence of an immediate sensory input. Such states, too, seem to occur in orderly sequence—at least those in logical reasoning do. It is not by chance that we speak of "trains" of thought. Moreover, when we try to transmit to somebody else this sort of information, we tend to do so step by step, in a concatenated series. Indeed, all our language communications—be they in English, Chinese, mathematical symbols, or clicks or grunts—are based on systematic, ordered information sequences.

Thus, a good part of our consciousness seems to be based on an orderly streaming of information states—a stream broken only during our sleeping hours and sometimes not even then. It is this constant streaming that probably gives us a sense of the flow of time.

Inner Time versus Physics Time

This much for the flow of time. But what of its direction? Why the asymmetry, the never-deviating progression from past to future? Offhand, there seems to be no intrinsic reason why that flow should always have the direction it has. Indeed, from the point of view of physics, it might as well go the other way.

Take classical Newtonian physics, for example. There the time t describing the evolution of a system has no preferred direction—it not only flows forward, as in our stream of consciousness, but also can flow backward. The laws governing a clockwork going in reverse are the same as those of a clockwork going the ususal way—the variable merely becomes −t. Time in Newtonian physics is inherently symmetric, and no consideration of boundary condition has any significant bearing on this pervasive property.

The same may be said for all of phsyics. All its successful equations can be used as well in one direction as in the other. In Einstein's relativity, time doesn't even flow. It is interwoven with space into one fabric where time flows no more than space does. And if we artificially reverse the way things normally unfold, say, we run a film of planets and stars backward, their movement still would conform to Einstein's laws.

All this is a far cry from the way we perceive time in our conscious experience and makes one wonder why we feel time as something always progressing in one direction. The Queen had it right when she summed things up for Alice and explained why people in Wonderland remembered things before they happened: "It's a poor sort of memory that only works backwards" (fig. 8.1). However, if you find that not cogent enough, I'll gladly back the Queen up with Hamilton's and Schrödinger's equations.

So, rather then laughing the Queen's argument out of court, we'd better find out what it is that makes it seem so absurd to us. Well, absurd, witty, waggish, or whatever you may call it, it is of the same ilk as that of a movie run backward. Such things never fail to tickle; and it's not the human factor alone that makes us laugh. The trick with the film works just the same with something as ho-hum as a stone falling into a puddle. As the stone lifts off from the water and is hustled by hundreds of amazingly collaborative water droplets, we instantly see that something is utterly wrong: it clashes with our notion of cause and effect—a concept borne of myriads of sensory experiences.

But besides the tail wagging the dog, something else here strikes us immediately as wrong. The scene clashes with our notion about the past and future, or, fully spelled out: *that we can affect the future but not the past.* This noesis is more deep-seated than the previous one. It has the same sensory roots, but it reaches farther back in evolution, perhaps into the ancestry of the human species. It may not always rise to the conscious surface, but we proclaim this notion in a thousand and one ways—we feel regret, remorse, and grief; we have expectations, aspirations, and hopes; it imbues our every mien, deed, and thought. That in principle things could be the other way around, that well-grounded physics laws insist that the future could become past, is of no

Alice: "I don't understand you. It's dreadfully confusing!"

Queen: "That's the effect of living backwards;
             it always makes one a little giddy at first -- "

Alice: "Living backwards? I never heard of such a thing!"

Queen: " -- but there's one great advantage in it,
             that one's memory works both ways."

Alice: "I'm sure mine only works one way,
             I can't remember things before they happen."

Queen: "It's a poor sort of memory that only works
             backwards."

Figure 8.1 Speculation on the backward flow of time.

immediate consequence to our conscious selves. To us, moments, minutes, hours, and even years have fled away into the past, never to be
recovered.

And that is the way of the world to us, *our reality . . . our time*, and
no amount of physicizing will alter that. If this looks like a contrarian
stance, it is only because it raises old ghosts. The question of time has

long been a source of dispute between those trained in biology and physics. The two sides, in the past, have instinctively formed camps. Generations of biologists went from cradle to grave growling about something rotten in the state of Physics—or, at the height of bonhomie, that something is missing. Well, it is true that the book of physics is still incomplete, but not in *this* regard; there isn't even a real bone of contention here. The whole polemic is no more than a balloon that goes poof at the first prick of Poincaré's mathematics; over any evolutionary significant period, time in the physics macrodomain flies by just as irretrievably and unidirectionally as in our sensory experience, as we shall see.

This brings me to the thermodynamics arrow, though perhaps too anthropocentrically for a physicist's taste. But after all, it wasn't a physicist but an engineer of machines designed for human comfort (Carnot) who originally came up with that arrow (and its law). This arrow points in the direction the events in the molecular sphere normally unfold. That's the sphere we are interested in here, the domain where life unfolds. Things in the molecular domain have not just an ordering in time but also a direction: from high to low order. And this trend prevails in the entire coarse-grained universe—the *macrodomain* of physics. We are no strangers to this trend: cookies crumble but don't reassemble; eggs scramble in the frying pan but don't jump back into their shells; chinaware shatters, waves fizzle, trees fall to dust . . . and all the king's horses and all the king's men couldn't put them together again.

The plight of the Humpty Dumpties of this world is independent of the universe's expansion (another arrow, the so-called astronomic one); in a contracting universe, the thermodynamics arrow would still point the way it does (it would do so even at the enormous spacetime warpage of black holes). The arrow represents a seemingly ineluctable trend and encompasses the entire sphere of molecules. The systems there constitute universes by themselves, which can harbor immense numbers of units—a drop of seawater contains some 10^{19} molecules of sodium salt. Those molecules dance perennially to the thermals and tend to get scattered randomly about (their equilibrium state). However, before they reach that higgledy-piggledy state, they show a penchant to deploy themselves in an orderly way, forming aggregates—loose molecular associations, tightly bonded compounds, and so on—the sort of molecular states chemists like for their concoctions.

All those states are rather short-lived. How much order a molecular system can embed depends on how much information there was to start with and how much gets pumped in afresh. But information is something hard to keep captive. It gives you the slip, as it were, and eventually will wriggle out of the strongest molecular bond—not even diamonds, which are all bonds, last forever. There is, in truth, no shackle, chain, or wall that could stop information from ultimately escaping—the thickest wall on earth won't stop the transfer of heat or shield off the gravitational force. So, in the long run, a molecular system left to its own devices becomes disordered.

That goes for complex systems too, including us. Our own organism holds immense amounts of information, and its systems are complex and tangled; but, made of molecules as they are, their eventual decay to disorder is inevitable. We, and the other living beings, manage to stay that fate for a while by taking in fresh information—or perhaps I should have said devouring, for we pump in information ceaselessly to stay alive. Alas, time's arrow points its pitiless course day in, day out, and nobody can pump enough to keep the horseman with the scythe away forever.

This much for time directionality and the distinction between past and future. As for the present, I intentionally left it out. Time flies continuously from being past to being future, and in such a continuum there is no room for a present. A scientist, thus, is entitled to give it the go-by. However, I do not wish to belittle something that is so deeply felt by everyone. I leave this theme, and not without a little envy, to someone like the writer Luis Borges. He knows how to celebrate life so brilliantly in the present and manages to soften even a diehard scientist, as he breathes life into Spinoza's philosophy of universal being (for which scientists have always had a weakness) and expounds that the wish of things is to continue being what they are—that the stone wishes to be stone, the tiger, tiger . . . and Borges, Borges. Well, for us scientists, that "being" is but a "becoming"—though we are perhaps the poorer for it.

As for the reasons behind time's arrow, its thermodynamics roots, these lie in the world of molecules, the world out of which life has evolved. There, fickle Chance holds court, and systems are tossed together by the throw of the dice. Molecular systems—their component states, the molecular aggregates—are statistical in makeup. However, this doesn't mean that these states are totally unpredictable. They are not as predictable as those of the stars and planets in the celestial sphere or those of the rise and fall of tides and apples—not with the same degree of certainty. The movements of these large objects are governed by Newton's laws, and his $F = ma$ predicts such events completely—which is not the case of a molecular system. Nevertheless, if enough information *is* on hand, one can still pin down the states of such a system with reasonable accuracy. Given some information about the significant variables in the molecular throw of dice, like position and speed, temperature and pressure, or chemical constitution and reaction rate, Newton's formula still works its magic. One has to bring in statistics then to cope with the vast numbers of molecular pieces. But this is not too difficult—that sort of statistical mechanics in the gross is rather routine nowadays (thanks to the spadework the physicists James Clerk Maxwell and Ludwig Boltzmann did in the last century). Unavoidably, one loses some definition here—that's in the nature of the statistical beast. But one brings home the bacon: the probability of the system's states. And when that probability is high enough, it still can claim the dignity of being a law of nature.

Thus, despite the turmoil and hazard of the die, we can still discern the regularities and predict events in the molecular world—the blurredness vanishes in the focus of the mathematics, as it were. There is a deficit of information in the molecular sphere, but that deficit is often small enough, so that *the future is determined by the past in terms of statistical probabilities*.

Now, this statistical future is precisely what the thermodynamics arrow chalks out. And given enough information we can descry the various underlying statistical events. But the overall direction of those events, the arrow's direction, we see intuitively without mathematics, as we know the ultimate state: complete disorder (the equilibrium state where the molecules just ride the thermals and move randomly about). So, in a bird's-eye view, the thermodynamics arrow of a system always points from the less probable state of order to the more probable state of disorder.

A loose analogy will make this clear. Consider a box containing a jigsaw puzzle whose many individual pieces have all been neatly put into place. If you shake the box, the pieces will become more and more disordered. Here and there, some of them may still hang together or fall into place, constituting parts of the original picture; but the more you shake, the more the pieces get jumbled up—it's the hand of time and chance.

There is no great secret to this hand. It is just the way things statistically pan out because there are so many more ways for the jigsaw pieces to be disordered than ordered. Just weigh the collective states of order available in the box against those of disorder: there is only one state where all the pieces fit together, but there are so many more where they don't. So it is highly probable for such a closed system to become disordered; and the more pieces the system contains, the more probable this becomes. For (closed) molecular systems, where the pieces number zillions, that probability gets to be so overwhelming that it constitutes a dependable rule: the second law of thermodynamics.

The reasons behind this run of things were discovered by Boltzmann. He derived the second law from first principles, namely, by tracking down the dynamics of molecular systems through a combination of statistics and Newtonian mechanics. His article appeared in 1877, but it has lost none of its luster and tells us more about time than a thousand clocks (Boltzmann 1877).

First and foremost it shows that molecular systems evolve toward states with less order—and why they do: because, as in the example of the jigsaw puzzle, there are so many more states with less order than with higher order. But between the lines, there is something that may, even a hundred years after Boltzmann, strike one as bogglingly surreal: that evolution can, in principle, also go the other way around. Indeed, that point, namely, full time reversibility, is implicit in the derivation, as it started from the time-reversible Newton's laws. But the proof is in the pudding, a theorem in this case. And that came not long after Boltzmann's coup, when the mathematician Henri Poincaré incontrovertibly showed that any system obeying Newton's laws must, in due time, return to its original state (or near it).

Thus, time's arrow *is* reversible—one can't argue with a theorem. And yet . . . and yet, doubt gnaws: Why don't we ever see time reversing? It's certainly not for want of wishing . How many dreams have been dreamed of reliving the past? What would we not give to turn the clock back! Here is one of those instances where intuition is likely to fail us. Our intuitive notions by and large get shaped by our senses, that is, by the sensory-information input into the brain—all our familiar notions originate that way, including those of time and space. But that very familiarity also is our weak point. It somehow perversely makes us blind to the broader realities; those notions get so dyed-in-the-wool from birth that we come to believe that they constitute the only possible way of representing the world. Only when we come to think about it do we see that they reflect only a limited information horizon.

The cold light of reality comes in from Poincaré's mathematics. These clue one in to the statistical probability for a time reversal, which is abysmally dismal for anything in the coarse-grained universe. One would have to wait an immensely long time before a reversal ever occurs—unimaginably long even by cosmological standards.

You can work this out yourself. The times for such a reversal (a "Poincaré's recurrence") go up exponentially with the number of molecules in a system: 10^N seconds for a system of N molecules. Even for a simple system, like a drop of saltwater, this comes to $10^{10,000,000,000,000,000,000}$ seconds, or about $10^{1,000,000,000,000}$ years, while the time since the big bang is a mere 10^{17} seconds or about 10^{10} years.

Information Arrows

Thus, time's arrow, though temporal in an absolute sense, is for all practical purposes irreversible. And under "practical" here I include time spans as long as those of biological or even cosmological evolution. Of arrows of so long a haul, however, we have hardly any direct experience (the arrows of our sensory world are only minuscule segments of those long-haul ones). Therefore, we have no option but to go beyond our natural sensory horizon, if we are ever to hope to bring those arrows into our grasp.

Such transcending would have seemed hopeless not long ago, as it would inevitably have landed us in metaphysics. Fortunately, however, such an approach not only is scientifically feasible nowadays but bright with promise. Thanks to modern technology, we now can search out the far reaches of space—the entire starry canopy above is at our beck and call here. What twinkles up there is a rich mine of time, and it doesn't take much quarrying to find what we need: we see the nearest star up there, as it was four years ago; the stars at the edge of our Milky Way, as they were 100,000 years ago; and the farthest galaxies, as they were 10 billion years ago.

So we can take our pick among the celestial arrows. Take that trusted sentinel of the north sky, the North Star, for example. This star, like any hot object, releases energy, and the direction of that energy flow is radially outward. This is what the word "radiation" stands for. But it also stands for the direction of the information flow, which is what matters to our brain—it isn't energy per se but information that makes us aware of the star. Energy is involved, of course, but only insofar as it embodies information and gets it down to us. The energy particles that serve as the carriers here are photons of 400–700 nanometer wavelengths. They stream through space and hit our eyes and transfer their package of information . . . *after* some time. Thus, the star's very existence implies an arrow pointing from the past.

That much we see with the naked eye. But with instruments, we can go one better. We then see the arrow not only pointing but beating time. Consider a neutron star. This is a star that has run out of nuclear fuel and contracted under its own gravity, collapsing to a body that typically is only a few miles across. But it is enormously heavy—it consists almost entirely of neutrons. And it spins fast. It makes hundreds of rotations per second about its polar axis, generating an intense magnetic field, at the same time that it spews out gamma rays, X-rays, and radio waves. These wavelengths we cannot see; they are beyond the capability of our visual receptors. But there are appropriate instruments in the scientific arsenal to make the stars visible.

That they are not directly visible actually has its advantage because we are not as easily led down the garden path as we are by the familiar verbal imagery that goes with our senses (a linguistic cross we bear from birth). The instruments in the scientific arsenal will receive the information from these neutron stars for us and convert it into a form decodable by our brain. Consider the longest waves, the radio waves. These get transmitted through the star's magnetic field as coherent beams. So, despite the distance intervening, such radio signals are heard loud and clear on Earth. They come in pulses of extraordinary regularity—which at one time mystified astronomers and gave rise to all sorts of science fiction about "signals from civilizations in outer space." Well, those signals are from outer space all right, but they are no longer a mystery. They result from the turning of the star; with each turn, the star's radio beam, like that of a lighthouse, sweeps past the Earth. And because the star is so heavy, it turns around at a steady speed; hence the extraordinary regularity of the pulses—the turning is constant within a few parts per quadrillion (15 decimal places!). Thus, that whirligig provides us with something more precious, though no less staggering, than the most extravagant fiction: a precise clock of cosmic time.

So everything on heaven's vault points an information arrow and beats the time. The arrows hailing from out there are long—some have been on the fly for 15 billion years. Those are the information lines issuing from the initial information state in the universe. Eventually, that initial state led, in the course of the universe's expansion, to condensations of matter and the formation of galaxies locally; and as these vast structures evolved, more and more structures—stars, planets, moons, and so on—formed locally inside them.

From our perch in the universe, we ordinarily get to glimpse only segments of these information arrows—local arrowlets, we might say. We easily loose sight, therefore, of the continuity. But as we wing ourselves high enough, we see that those arrowlets get handed down from system to system—from galaxy to stars to planets . . . to us.

A Very Special Arrow

In an all-embracing view, then, information arrows radiate out from the origin to the four corners of the universe (fig. 8.2). They nurse all organization. Their primary destinations are the galaxies, the organizational units in the universe; but each arrow has innumerable branches whose prongs reach into every nook and cranny of those units. Together, arrows, arrowlets, and prongs constitute a branchwork of staggering proportions—so staggering that one gratefully seizes on any expedient enabling one to simplify so complex a reality. Thus, making the most of their common directional sign, we lump them together.

Well, except one. This one we could hardly bring ourselves to toss in with the lot. It bears for coordinate 0,0,0 . . . and directly down on us. This arrow, willy-nilly, takes the limelight. Though, over a stretch of its flight, it is not as clear as we would wish it to be. A significant portion of the information in that stretch—the leg from the origin to the galactic unit—is carried by gravitons, the physics of which is still in a state of querulous unease. But the last leg, from the Sun to us, is straightforward. There, the bulk of the information is carried by the dependable photons. These particular ones have wavelengths between 300 and 800 nanometers (from the ultraviolet to the infrared), and our planet gets liberally showered with them—260,000 calories for each square centimeter of the Earth. Which amounts to 26×10^{24} bits of information per second, an enormous information flow. And this is the stream from which we suckle—we and all organizations of life on Earth.

This stream gets into the living mass through a window formed by a handful of simple molecules: the green chlorophylls and yellow β-carotenes of plants and bacteria, plus a few other ones of this sort, which are red and blue. The pigments are quite pretty, and their color catches our eye. But what really matters here lies hidden inside the molecules' electron clouds. There, a set of electrons—strategically deployed around the carbon chains and rings—are critically poised for the photon energy quantum. These electrons lie in wait for the photons, so to speak; and when one comes along, they devour it, body and soul (fig. 8.3).

The molecules here are of modest size—chlorophyll has only a few dozen atoms, and β-carotene, even less. They also are structurally quite simple. β-Carotene, for example, consists of just a single carbon chain, with 11 electron pairs forming the photon trap; and chlorophyll has a somewhat larger carbon-ring structure stabilized by a magnesium atom. In information terms, these organic molecules don't amount to much;

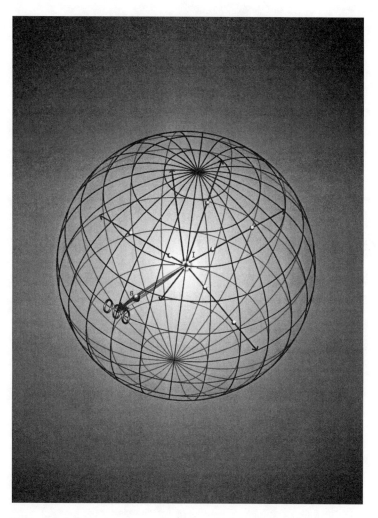

Figure 8.2 Information arrows. Radiating out from the initial information state (*I*), the arrows bear for coordinates where organizations take place in the expanding universe. The dots on the arrows mark major way stations—galaxies, stars, planets, and so on; *S* is our Sun. Out of the countless arrows in today's universe, only a handful are diagramed (in bold, the two bearing down on coordinate 0,0,0).

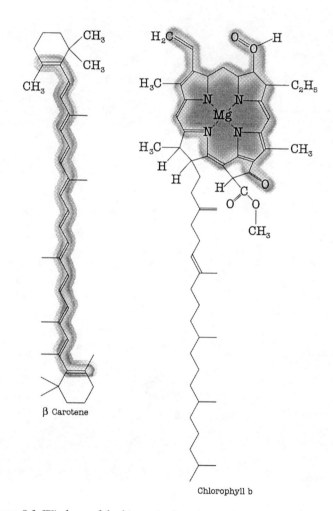

β Carotene

Chlorophyll b

Figure 8.3 Windows of the biomass—the primary photon traps. The chemical structures of two, chlorophyll b and β-carotene, are shown here; the photon-trapping loci (the regions with alternating single and double bonds) are shaded. These molecules are tuned to photon wavelengths between 400 and 500 nanometers—between the violet and blue/green. (The rest of the visible spectrum is covered by the other molecular traps of this sort, which are not shown.)

and we wouldn't expect otherwise, because they go back to primeval times—some of them to more than three billion years—when there was still preciously little biological information around. They are the results of a long series of trial and error in the laboratory of Evolution, the products of a search among the primal organic molecules for a good energy match with the bountiful photons.

We now also catch on to the other side of Evolution's game, the photons' side. The visible photons are the only ones whose quantum is compatible with the energy level of the covalent bond, the electromagnetic energy that holds the atoms of organic matter together. The photons of higher quanta, like the ultraviolet (though well absorbed by organic molecules and probably as abundant in the primeval atmosphere as the visible photons) break the bonds apart; and those of lower quantum, like the infrared and radio waves, are taken up randomly and directly converted to thermal energy modes.

And this is how it came to be that these tiny specks of energy and matter play such an important role on Earth. They are the Chosen Ones. They are masterpieces of harmony between the cosmic and the terrene.

The Demons behind the Arrow

Now, as to the information in the solar photons, this doesn't just dissolve and melt away in the biomass. Rather, it enjoys a long and eventful afterlife: after being trapped, the information gets first stored in chemical form and then trickled down along molecular chains to the places where higher molecular organizations occur. This is the stomping ground of biochemists and molecular biologists—and their delight. But when a physicist barges in, what piques his interest is that this information seems to be forever on the move—and forward. This is all the more astonishing because the movement is based on the random walk of molecules. By all rights, a system like that should be very noisy—molecular random walk is inherently so. Thus, everything here would seem to militate against the information making headway against the noise. Yet, despite the odds, it does get through. Why?

This question strikes at the heart of life. Short of flouting the laws of physics, there seems a priori to be only one possible answer: The information in living sytems is optimally encoded. In fact, this turns out to be the case. But let me first explain what I mean by "optimally encoded."

That terse qualifier stands for a mode of encoding in information theory that overrides the communication noise. All modern communication media use such encoding. Indeed, they depend on it—without it, their signals would be drowned by noise. This bit of savvy, however, didn't come from technology but from a theorem; and when this theorem first appeared, it took everybody by surprise. It was the brainchild of the mathematician Claude Shannon. Musing about the noise bedeviling all communication channels, Shannon proved that noise does not, by itself, set a limit to information transmission (Shannon 1949). He showed—and this was like a bolt out of the blue—that the errors in information transmission due to random noise can be made arbitrarily small by proper encoding. Specifically, that in any digital information transmission the probability of error per bit can be made arbitrarily small for any binary transmission rate. In plain language: when one can't get rid of the noise in communication—and this holds for any communication channel with finite memory—*one still can transmit a message without significant error and without loss in speed of transmission if the message has been properly encoded at the source.*

This was one of the great triumphs of information theory. The theorem had a number of immediate practical spinoffs. Within just a few years of Shannon letting the cat out of the bag (1949), wire- and wireless lines cropped up everywhere, buzzing with encoded digitized information. Shannon, we might say, single-handedly ushered in the age of global communication.

But little did he know that Lady Evolution had beaten him by aeons. She came up with optimal coding formulas long before humans were around, even long before multicellular organisms were. Her search for such formulas, we may assume, started about four and a half billion years ago, as soon as she chose water-soluble molecules as information carriers. Such molecules offered vehicles for transporting information at a bargain; they could just ride the thermals in the water. But that mode of transport was noisy; so, for Evolution, coding was a Hobson's choice.

Indeed, the search for adequate molecular codes took up a good part of her available time. The products of that search are now everywhere around and inside us. They are modular structures, macromolecules made of nucleotide or amino-acid modules, ranging in size from a few hundred modules (the oldest vintage about four billion years) to thousands of modules (the newest a few million years), large enough to be seen in the light microscope.

But what makes such random-walking signals reach the end of the communication line? It is all very well and good that they are well encoded; but what gives them a direction, and how do they get through the line? The first question has an easy answer. The direction is given by the molecules' concentration gradients, which, thanks to amplification along the line (largely gene amplification), are steep enough in all living organisms to point an arrow. The getting-through is another matter. Things, in this regard, are decidedly not as simple as in a telephone line. There, all you have to do is impress a gradient (a voltage), and the information carriers (the electrons) will run down the length of the wire. In the biological lines, the information gets relayed from one carrier to another—one particular molecule in a sea of molecules transfers information to another upon random encounter.

Now, such haphazardous communication requires that these molecules somehow recognize each other. And this has its price—there is no free lunch in cognition. The laws of thermodynamics demand that someone here bring in information from outside to settle the entropy accounts. That "someone" is a molecule—it has to be; the system is autarchic. Indeed, a member of the encoded macromolecular clique draws the requisite outside information onto itself and funnels it into the communication lines.

This is a talent nowhere to be found outside the orb of life. All molecules, including the inorganic ones, to be sure, can carry information—their intrinsic one. But few even among the macromolecules can summon the *extrinsic* information needed for an entropy quid pro quo. The simplest macromolecules with that talent are made of nucleotide modules—oddly metamorphic RNAs, made of only a few hundred modules, which can switch from one-dimensional to three-dimensional form. Those molecules came early onto the evolutionary scene, and they were probably the first to be up to the task (they can still be found at the lower zoological rungs and in our own ribosomes). But the predominant ones today are proteins. These are made of thousands of amino-acid modules complexly deployed in three dimensions; and what is more, the modules can move with acoustical speed, causing the molecule to quickly change gestalt.

Macromolecules of this sort arrived rather late on the biological scene (much later than the tiny photon traps). But they have the entropy thing down pat. Not only do they procure the requisite information but they mete it out precisely to where it's needed in the communication lines. Indeed, so prodigious are their powers that they are in a class by themselves: cognitive entities.

One is not used to thinking of molecules as having smarts, certainly not the molecules one usually deals with in physics. One physicist, though, would have felt at ease with them: James Clerk Maxwell, the towering figure of the past century who, among other things, gave us the electromagnetic theory. In Maxwell's time, macromolecules were not known, let alone the cognitive ones; but had they been, he would not have hung in doubt about what to call them. In 1871 he published a famous gedankenexperiment which was to nurse the curiosity of physicists for over a hundred years (Maxwell [1871] 1880). He toyed then with a hypothetical sentient being—a "demon," in his words—who could see molecules and follow them in their whimsical course. His demon was an artful dodger of the second law who could distinguish individual molecules and pick them out from a gaseous mixture, creating order from chaos.

Well, the proteins here have exactly the talent of Maxwell's demon: *They can pick a molecule out of a universe.* But there is this difference: *They do so entirely within the law.* Their talent is built right into their three-dimensional structure. They have a cavity that fits the contours of that particular molecule, and they hold fast to it through electrostatic interaction. The electromagnetic forces involved here operate only over very short distances, 0.3–0.4 nanometers. So only a molecule that closely matches the cavity's shape will stay in place—one molecule out of a myriad.

But such electromagnetic finesse is just a part of the trick. The most exciting part takes place deep inside the protein molecule, and it takes the information lens to spot it. The demon can't get around paying the thermodynamic price for his cognition. Nobody can. He does so at the end of the act by extracting the requisite negative entropy from organic phosphate—adenosine triphosphate (ATP) or the like. Such phosphates are chemical storage forms of solar energy concentrated in the phosphate tetrahedral bonds. The demon, with the aid of water, manages to break those bonds and set their energy free. But it is not the energy he hankers after (though you may easily be led to think so from reading biochemistry books). There is a good deal of energy stored in these phosphates, it is true; and, as the tetrahedra break apart, all this goes off in one blast (7.3 kilocalories per mole). But that's just flimflam. Much of that energy goes down the drain (it merely heats the environment). Indeed, all of it would, if the demon weren't there and extracted his bits from it.

Thus, the demon's stratagem stands revealed: He nabs information that otherwise would go to waste—that's why I called it "negative entropy" earlier. His trick is—and it is easily the most dazzling of magician's acts—to wrest information from entropy!

Let us now take a closer look at the demon's sleight-of-hand. It isn't easy—he juggles fast. He twists and wriggles as his amino-acid modules swivel, causing his molecular configuration to change in a fraction of a second. But if you watch closely, you see that for all his writhing, he just switches between two gestalts: in one, his cavity is open; and in the other, it is not. He methodically goes through a cycle. In gestalt 1, he is ready to catch the object of his desire. But no sooner does he score than he shifts to gestalt 2 and bids his mate goodbye. Then he reverts to gestalt 1, ready for more of the same (fig. 8.4).

But not before he settles the thermodynamics accounts. And he pays cash on the barrelhead—there are no such things as installment plans in the thermodynamic universe. Indeed, it is precisely when he reverts to his original gestalt that he pays the entropy price, the bits he palmed from the phosphate.

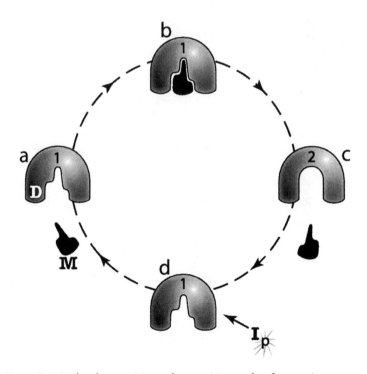

Figure 8.4 Molecular cognition—the cognition cycle of a protein demon. The demon (D) goes through a cycle (a → b → c → d → a), switching between gestalts (1,2). The electromagnetic interaction with the molecular mate (M) triggers the switching from gestalt 1 to 2, and the uptake of information from phosphate (I_p) allows the structure to switch back to gestalt 1.

There is an intriguing analogy here with computers. The information states of a digital computer (its "logical states") are embodied by distinct configurations in the hardware, just as the information states of the protein are embodied by distinct molecular configurations. Moreover, the computer hardware gets reset when the memory register gets cleared at the end of the cycle. The way this shapes up with computers has been penetratingly analyzed by the mathematicians Rolf Landauer (1961) and Charles Bennett (1982). They showed that the crucial settling of thermodynamics accounts takes place as the memory is cleared.

And so it does here in the protein molecule. In both the protein and the computer, the critical extrinsic information comes in during the resetting of the structure. It is then that the demon pays the thermodynamics piper. (My book *The Touchstone of Life* gives a full account of this.)

A protein demon typically goes through hundreds or thousands of cognition rounds. And at the end of each, he antes up the mandated information. He draws that information in from outside—he and countless other demons in an organism. Thus, in the aggregate, those demons bring in a gargantuan amount, a sum by far exceeding the amount coming from the DNA. This is all too easily overlooked, as one naturally tends to gravitate to the DNA, the spectacular repository of organismic information. However, that is only static information— information in two-dimensional storage form. To get such information to move, you need an altogether different kettle of fish: something that can deploy information in four dimensions. This is a tall order to fill because, apart from the dimensional problem, the deployment has a tremendous thermodynamic cost. But the macromolecular demons have what it takes: the information, the three-dimensional agility, and the wherewithal to defray the cost.

So now we know who shoulders the arrow at coordinate 0,0,0. We now also can give it a proper name. That arrow has, over some three billion years, been engendering forms of ever-increasing complexity and order. I use "engendering" (for lack of a better term) in the sense of *immanent* cause, so that the word has an etiological connotation rather than the ordinary lexical one of "generating," an action that cannot beg force. In terms of information theory, it is as correct to say that order is inherent in information as the more familiar reverse. So we'll call it *Life's Arrow*.

But my purpose here was not to unfurl the flag; it was to show the continuity of the information arrow and the evolutionary continuity between the cosmic and biological. It is hard to say where the one ends and the other begins. This is the nature of the evolutionary beast. But, for the sake of argument, let's assign the beginning of the biological part to the emergence of the small photon-trapping pigment molecules. Others may prefer to make their count of biological time from the emergence of molecular demons. On such long scales, it hardly matters from which of these points we start. Either way we get about four billion years. And onward, if all goes well, another four billion years are in the offing. By then our sun will have burned out, and when that photon source dries up, there is no help for it, life's arrow will cease on this planet.

A Second Arrow

I end my story with a sketch of a second arrow. This arrow comes from the same bow, but it had to wait in the wings, as it were, until conditions were ripe on coordinate 0,0,0 for a change of mode in information transfer. The ground in some locales had to be prepared for molecular information flow to shift from three dimensions to one dimension. That took a while—some three billion years; it is no simple thing to change from free and easy wanderlust to goose step.

The turning point happened some time between 700 million and one billion years ago, when certain cells grew sprouts (dendrites, axons) that were capable of conducting electrical signals down their length. These were digital signals—brief (millisecond) pulses of invariant voltage—that were transmitted at speeds of the order of 1 meter per second (and eventually up to 100 meters per second) several orders of magnitude higher than the signals in the ancient web; moreover, the signals were strong (a tenth of a volt)—well above background noise. Given this uptrend, it didn't take long for multicelled organisms to become crisscrossed with new and faster lines of communication. Indeed, a whole new web sprang up—the neuronal web—which used the digital signals for communication *and* computation. Thus, a sector of the biomass soon began to squirm and flutter . . . and steal the spotlight.

The second arrow is shouldered by demons, too; but these particular ones all dwell in cell membranes rather than the cell interiors, as many of the demons ministering to the first arrow do. Their function may be gleaned from mere externals: they are large and made of several proteins units that are symmetrically deployed about a central axis, leaving a narrow water channel in the middle (fig. 8.5). These structures are ostensibly designed to funnel small inorganic molecules (ions) through cell membranes. But there is more to this than meets the eye; and, once again, it takes the information loupe to spot the cloven hoof:

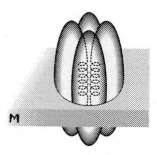

Figure 8.5 Diagram of a membrane channel made of four protein units. The helical stretches from two units forming the channel lining are sketched in; they constitute an electrostatic shield for small ions. (*M*) is the cell membrane. (From Lowenstein, *The Touchstone of Life*, reproduced with permission of Oxford University Press and Penguin Books.)

an electrostatic shield (a constitutive part of the α-helical channel lining) that allows the proteins to pick certain ion passengers, or a certain class of ion passengers, out of the crowd.

This modality of cognition is no great shakes (in information power, it ranks well below the cognition by the protein demons we saw earlier). But what makes these proteins truly formidable is their partnership with other membrane proteins. This is a close functional partnership and sometimes even a structural one. The protein partners here are consummate demons. They are capable of picking out certain pieces of information from outside the cells —information from molecules or energy fields. And conjointly with their channel partners, they convert it to electrical form and spread it over the cell membrane. In higher organisms, that partnership occurs in neurons, namely, on the tips of dendrites facing the world outside, or in specialized cells attached to neurons. This is the nub of our senses where that information flows in staccato form via dendrites or axons to the brain.

Let's put these sensory demons in an evolutionary context. They form a whole coterie—its various members came on the evolutionary scene at different times. I will pick out one who arrived about 700 million years ago, rather late by evolutionary standards. This specimen inhabits our eyes and goes under the name of "rhodopsin"—a colorful protein with a photon trap at the center, not unlike β-carotene.

This newcomer is sophisticated. His antennas are directed to the quantum realm—not the molecular world the antennas of all other sensory demons are directed to. He is tuned to photons in the range of

400 to 700 nanometers wavelength (the range from red to blue). And he is as sensitive as he possibly could be: to one photon quantum! He is able to raise that minimum of energy up to the macroscopic realm and, jointly with his channel partner in the sensory-cell membrane, to convert it into an electrical signal. It is in this form that the photon information gets into the brain where it gives rise to a visual sensation—a state of consciousness.

Thus, this demon is the window through which the second arrow enters—our window to the world without time.

To conclude, I allow myself to go outside my bailiwick for a general remark. I am encouraged to do so by the presence of Sir John Templeton at the symposium and by the counsel he offers in his book (2000). I hope I managed to provide a little taste of the power that information theory wields in our probings of life. It gives one a wondrous strange feeling to be able to read the script that nature renders and to peer into her designs. This makes one confident as a scientist and not a little primped by human pride. One usually calls such peering, when successful, insights, and I have done so myself quite a few times. But in the spirit of a Templeton Symposium, I will make my mea culpa here: from a higher perspective, such insights are but the benefits of four billion years of hindsight. Yet it is a way, and for a scientist the most satisfying one, to see God in his splendor. I can put it in no other way.

REFERENCES

Augustine of Hippo. [c. 400] 1991. *Confessions.* Translated by H. Chadwick. Oxford: Oxford University Press.

Bennett, C. H. 1982. "The Thermodynamics of Computation." *International Journal of Theoretical Physics* 21: 905–940.

Bennett, C. H. 1987. "Demons, Engines and the Second Law." *Scientific American.* November: 108–116.

Boltzmann, L. 1877. "Über die Beziehung eines allgemeinen mechanischen Satzes zum zweiten Hauptsatze der Wärmetheorie." *Akademie der Wissenschafte, Wien, Mathematisch-Naturwissenschaftliche Klasse, Sitzungsberichte* 75: 67–73.

Boltzmann, L. 1895. "On Certain Questions of the Theory of Gases," *Nature* 51: 413–415.

Davies, P. C. W. 1974. *The Physics of Time Asymmetry.* Berkeley: University of California Press.

Davies, P. C. W. 1995. *About Time: Einstein's Unfinished Revolution.* New York: Simon and Schuster.

Halliwell, J. J., J. Perez-Mercader, and W. H. Zurek. 1994 *The Physical Origins of Time Asymmetry*. Cambridge: Cambridge University Press.

Landauer, R. 1961. "Irreversibility and Heat Generation in the Computing Process." *IBM Journal of Research and Development* 5: 183–191.

Loewenstein, W. R. 1961. "Biological Transducers." *Scientific American* (August) 98–108.

Loewenstein, W. R. 1965. "Facets of a Transducer Process." In *Sensory Receptors*, edited by M. Delbrück. Cold Spring Harbor Symposia on Quantitative Biology, no. 30: 29–43.

Loewenstein, W. R. 1999. *The Touchstone of Life. Molecular Information, Cell Communication, and the Foundations of Life*. New York: Oxford University Press.

Maxwell, J. C. [1871] 1880. *Theory of Heat*. 6th ed. New York: Appleton.

Shannon, C. E. 1949. "Communication in the Presence of Noise." *Proceedings of the Institute of Radio Engineers* 37: 10.

Templeton, J. 2000 *The Humble Approach in Theology and Science*. Philadelphia: Templeton Foundation Press.

PART III

Philosophical and
Religious Perspectives

NINE

EMERGENCE OF TRANSCENDENCE

Harold J. Morowitz

While thinking about this chapter, I happened to read the following remarks by Pope John Paul II and John H. Holland. The pope, in giving the allocution to the plenary session of the Pontifical Academy of Science of 1992, entitled "The Emergence of Complexity in Mathematics, Physics, Chemistry and Biology," asked, "How are we to reconcile the explanation of the world . . . with the recognition that 'the whole is more than the sum of its parts'?" (Pullmann, 1996, 466) Holland, in his book *Emergence: From Chaos to Order*, notes: "Emergence, in the sense used here, occurs only when the activities of the parts do *not* simply seem to give the activity of the whole. For emergence, the whole is indeed more than the sum of the parts" (1998, 14). Curiously, the identical words "the whole is more than the sum of the parts" occur in the scientific treatise and the papal discourse. This seems to speak eloquently to the spirit behind this volume. Within that spirit, I would like to address my remarks to emergence and reductionism, not as antagonistic approaches but as complementary aspects of understanding.

Reductionism is applicable to disciplines where the material of study may be organized in a hierarchical order. Reductionism is the attempt to understand structures and processes at one level in terms of the constructs and understanding at the next-lower hierarchical level. Emergence is the attempt to predict the structure and activities at one hierarchical level from an understanding of the properties of agents at the next-lower level. Reduction and emergence thus go in opposite directions in the hierarchy. Emergence is less deterministic, because if there are a large number of agents and the interaction rules are nonlinear, then combinatorics generate a number of possible trajectories too enormous for the largest and fastest conceivable computers. If the emergence of the world is to be other than totally random, then rules must be operative as to which trajectories are favored. Understanding these selection or pruning rules is a major part of the science of complexity.

The power of complexity and emergence is that the possibility then is open both to understand and accept the approach of reductionism and the novelty that comes with emergence at each hierarchical stage. The unpredictable novelty results from the combination of the reductionist rules with the selection rules and pruning algorithms. This provides a method to study how the whole is different from the sum of the parts. With respect to emergence, we follow Holland's caveat that "[w]hat understanding we do have is mostly through a catalog of instances" (1998, 3). Thus to explore the whole, we examine a chain of emergences from the primordium to the spirit. Then we move from the instances in the chain to look for the consequences of emergence and the meaning in a broader context.

The emergences I choose to focus on are

1. The primordium
2. Large-scale cosmological structures
3. Stars and nucleosynthesis
4. Elements and the periodic table
5. Solar systems
6. Planetary structure
7. Geospheres
8. Metabolism
9. Cells—prokaryotes
10. Cells—eukaryotes
11. Multicellularity
12. Neurons and animalness
13. Deuterostomes
14. Cephalization
15. Fish
16. Amphibians
17. Reptiles
18. Stem mammals
19. Arboreal mammals
20. Primates
21. Apes

22. Hominids
23. Tool makers
24. Language
25. Agriculture
26. Technology
27. Philosophy
28. The next emergence (The spirit)

While these instances are chosen to study emergence, when they are arrayed in the temporal sequence given here, they present a 50-year update of Teilhard de Chardin's (1961) unfolding in time of the evolution of the noosphere and the moving on to the next stage of the emergence of the spirit (in *The Phenomenon of Man*). It is in that spirit that we may look to the dialogue between science and theology.

The primordium and large-scale cosmological structure are part of the mystery of mysteries, the beginning. There are questions of how far back we can go, and Lee Smolin's idea, as outlined in *The Life of the Cosmos* (1997), is an attempt to make those questions part of normative science.

The emergence of stars introduces nucleosynthesis and a novelty in the system. To the two types of nuclei are added 90 others and a vast complexity now possible for the system. At this point a new principle kicks in: no two electrons can have the same quantum numbers. This pruning principle leads to the periodic table, chemical bonding, and structure formation. The more general formulation of the system is that all functions representing states of two electrons must be antisymmetric. These functions are the probability distributions of quantum mechanics, and the principle is designated the Pauli exclusion principle. All of chemistry follows from this principle. It is a selection rule because it admits only antisymmetric states and therefore vastly reduces the number of possible configurations. The exclusion, however, is nondynamical; it doesn't follow from force laws but rather has a kind of noetic feature. It occurs at a very deep level in physics and leads to the emergence of chemistry. The philosopher of science Henry Margenau has looked beyond this and notes:

> The advance has also led physics to higher ground from which new and unexpected approaches to foreign territory can be seen. Not too far ahead lies the field of biology with its problems of organization and function, and one is almost tempted to say that modern physics may hold the key to their solution. For it possesses in the Pauli principle a way of understanding why entities show in their togetherness laws of behavior different from the laws which govern them in isolation. (1977, 443)

Note that the theme of the whole being different from the sum of the parts noted by John Holland and Pope John Paul II is here seen an essential part of physics. As a matter of fact, Margenau notes, "[w]e can understand how the whole can be different from the sum of the parts" (445). He gives an example:

> In the process of constructing a crystal from its atomic parts, new properties are seen to emerge, and these properties have no meaning with reference to the individual parts: among others, ferromagnetism, optical anisotopy, electrical conductivity appear, all "cooperative phenomena" (the term is actually used in the theory of crystals) which owe their origin directly or indirectly to the exclusion principle. (445)

Margenau goes on to suggest that there may be other nondynamical principles at higher hierarchical levels from which other properties of the whole may emerge—perhaps life and mind.

Steps 5–8 deal with the emergences of solar systems and planets. The addition of chemical sophistication to celestial mechanics introduces a new level of emergence seen in the structures and processes studied in geochemistry and geophysics.

At this stage I want to step back and realize the stage that has been set by the unfolding of the laws of physics. If we follow the religion of Spinoza and Einstein, the unfolding universe is God's immanence or the immanent God. But emergence provides a new view. Within the universe novelty is possible, and clearly within our corner of the universe the major novelty is life, which is propaedeutic to all other novelties.

The emergence of protocells involves as agents the molecules of carbon, hydrogen, nitrogen, oxygen, phosphorus, and sulfur and as rules the reactions of organic chemistry and the phase behavior of amphiphiles. If we think of the network of all possible organic reactions as an information space, then the partitioning by amphiphiles connects information space with physical space. The origin of protocells involves a concentration of material in both spaces and this is driven by chemical rules. For example, in network space, whenever an autocatalytic subnetwork appears, it is a sink for carbon. When amphiphiles are synthesized, consisting of a polar head and hydrocarbon tail, then if the tails of the right chain length are present, they condense into planar bimolecular leaflets. These will condense into vesicles. If these encapsulate molecules of the information network, we are well on our way to protocells. Given a source of amphiphiles or the synthesis of amphiphiles in the network, then vesicles can grow and divide.

Given identifiable replicating entities, competition for resources is a necessary outcome, and fitness as a selection factor automatically enters the picture. That is, we are in a Darwinian domain from the very beginning. This led to stabilizing of information by templates and genetic molecules, and the prokaryotes emerged. They developed great chemical diversity and motility. With motility, primitive behavior emerged. For example, bacteria can, from a statistical point of view, swim uphill in a nutrient gradient and downhill in a toxicity gradient. This requires perception and response.

The next major emergence is eukaryote cells, which are the result of a new process, endosymbiosis, by which cells merge and one of the merged components evolves into an organelle. This process expands the information space within the cellular physical space. When multicellularity emerges, various components in the information space partition into physical spaces, and specialized cells appear on the scene.

The next series of emergences that are of interest are along the evolutionary tree of the animals. The governing rules of competition lead to niche selection, evolution of new forms, and environmental shifts creating new niches. From the present perspective, we are interested in the emergence of the animal mind.

The notion of the mind leads us to an epistemological paradox. From the point of view of the philosophy of science, we start with perception, and mind is the primitive and material objects are constructs. The point of view of emergence starts with the big bang and the distinguishable particles that form nucleosynthesis and asks how mind could have emerged in a complex series of steps starting from the properties of the Pauli exclusion principle. Thus mind constructs matter, and matter generates mind. This circularity is deep. It is somewhat reflected in the title of Gregory Bateson's book *Mind and Nature: A Necessary Unity* (1979). It is reflected in Margenau's argument that something akin to mind already manifests itself in the nondynamical consequences of the exclusion principle.

Psychologist Donald R. Griffin, writing in *Animal Minds* (1992), takes a less universal but nevertheless very broad view of mind, which he sees as extending through the animalia: His approach entails considering animals as conscious, mindful creatures with their own points of view, and he attempts to "infer, as far as the available evidence permits, what it is like to *be* an animal of a particular species under various conditions." He chooses invertebrate as well as vertebrate examples, with several chosen from the insects.

I have discussed behavioral responses to environmental gradients going back to examples from bacterial motility. This requires that individual bacteria sense the environment and temporal changes in the environment. The responses in these cases seem to be genetically preprogrammed.

One of the difficulties in all responses' being preprogrammed is that there are vastly more environmental states than genetic responses within a finite genome. Griffin then argues that mind, whatever it is, would have a fitness advantage and thus an evolutionary edge.

> As pointed out by the philosopher Karl Popper (1978), what he termed "mental powers" are presumably helpful to animals in coping with the challenges they face, and therefore must contribute to their evolutionary fitness. He emphasized how useful it is to think about alternative actions and their likely consequences before they are actually performed. This of course serves to replace trial and error in the real world, where error may be costly or fatal, with decisions based on thinking about what one may do. Popper seemed to imply that such mental trial and error is a uniquely human capability, but the versatility of much animal behavior suggests that on a very simple and elementary level they sometimes think about possible actions and choose those they believe will lead to desired results or avoid unpleasant ones. (22)

With this in mind I return to the next evolutionary emergence, the neuron. Before the evolution of neurons, most cell-to-cell communication was chemical, signaling molecules released by one cell were picked up by another cell, which responded. The time response was limited by diffusion speed, which is quite slow over distances of a few cell diameters.

Neurons receive a chemical signal at one end and convert it to an electrical signal, which very rapidly traverses the length of the cell, where signaling molecules are released. This was the indispensable invention for animals to become large and highly multicellular.

The next series of emergences are in the domain of the vast evolution—the radiation over the past 600 million years. It starts with primitive wormlike creatures with cephalization emerging as an organization plan. In a group of animals known as the deuterostomes, the sensory organs and the ingestive apparatus moved into the head region to provide a strongly axial character as well as an executive region, the head. The neural network became centered in the head, as distinguished from the distributed ganglions in a number of other animal types.

In thinking about purpose in the emerging world, there has been an impatience in viewing the way that the actual processes take place. After all, what the biblical chroniclers describe as a six-day process, we are describing as a 12-billion-year process, a scaling of 700 billion to one on the time scale. It seems arrogant in the extreme for us to be impatient about the rate at which divine immanence unfolds into the domain of complexity that make transcendence possible. The rate of complexification appears to be constantly increasing over the biological domain. From prokaryotes to eukaryotes took two billion years. Multicellularity took another billion years. Moving from sea to land took a few hundred million years.

The emergences from fish to amphibians to reptiles has to do with occupying niches from totally submerged to totally terrestrial. Accompanying this habitat change was progressive increase in brain size and development of limbs. There also appear to be changes in behavioral repertoire that add a new dimension of complexity.

Sometime about 100 million years ago, mammals evolved from the reptiles. Although there are many differences among these taxa, the process begins with a prolonged development of embryos in one of the parents with live birth and prolonged parental interaction. Parental offspring interaction also evolved among the birds.

What is of major importance with this feature of parent offspring interaction is that learned behavior enters into the fitness criteria and a kind of Lamarckianism enters into the evolutionary process. Lamarck hinted that acquired characteristics could be inherited, and in a physiological or morphological sense this was rather convincingly disproved by the separation of germ cells from the somatic cells of organisms. There was a radical and abrupt change in this doctrine with the training of offspring by parents. For a taxon now had, in addition to a pool of genetic information, a pool of behavioral information that could be much more rapidly changed. This was a major shift from the genetic to the noetic that has characterized succeeding evolution among the mammals.

There is an important principle in evolutionary biology that says that given an available niche, a given major taxon will evolve species to occupy that niche. Following the cretaceous tertiary transition when the mammals were undergoing an evolutionary radiation, major forests were developing around the world. Arboreal mammals evolved, including insectivores, fruit eaters, and carnivores. The arboreal mammals included bats, shrews, and rodents. Certain features such as grasping hands, stereoscopic vision, and a functional tail all contributed to fitness, and a class of prosimians emerged.

Within the primate grouping from the prosimian to the great ape there is a gradient of enlargement of the cerebral cortex. The higher primates involve societies of interacting known individuals.

The hominids appeared on the scene some six million years ago. The subsequent period is the social Lamarckian age as contrasted to the four-billion-year biological Darwinian age. This exemplifies the speed-up that has characterized the changing time scale of emergences. The first four million years of hominids with evidence of numerous *Australopithecus* species is clearly an age of hunter-gatherers living in communities of a relatively small number of individuals. It is the period of the emergence of *Homo* from hominid. We know very little about this age, other than the fossil remains; however, there is considerable knowledge of the great apes and of *Homo sapiens* to attempt to reconstruct how we came to be.

From about two million years ago, we begin to find remains of *Homo habilis*, his tools and other stone artifacts. We can reconstruct various evidence of *Homo erectus* and finally the various groups of *Homo sapiens*. From *Homo habilis* to modern man is only about 60,000 generations. Again, there is a progressive speed-up in the emergence of the features I wish to discuss.

Homo sapiens emerged about 200,000 years ago (6,000 generations). The transition from the hunter-gatherer stage to the age of agriculture took place about 10–15,000 years ago (400 generations). The age of settled *Homo sapiens* comes close to the biblical time domain. Urban centers have been around for about 5–7,000 years.

At some time in that settled period, the beginning of closing the loop emerged: mind thinking back on the beginnings and the development of mind. At this stage, ideas of god or gods entered into history.

Among the Hebrews, the God of History emerged. This extracorporeal God interacted in human affairs to influence the world of social groups, in particular God's people. This was a transcendent God who rarely communicated the divine will to prophets.

The Greeks developed two kinds of god: the superhumans of the state religion and the God of Plato and Aristotle, the unmoved mover who governed the laws of the universe independent of human will and desire. To construct an immanent God required other constructs: the laws of nature. Thus, the scientific concepts of the Ionian savants were a prelude to the immanent God, although there are hints of this idea in writings such as the book of Job (Job 38:4–8):

4. Where was thou when I laid the foundations of the earth? declare, if thou hast understanding.
5. Who hath laid the measures thereof, if you knowest? or who hath stretched the line upon it?

6. Whereupon are the foundations thereof fastened? or who laid the corner stone thereof;

7. When the morning stars sang together, and all the sons of God shouted for joy?

8. Or who shut up the sea with doors when it brake forth, as if it had issued out of the womb?

With post–Pauline Christianity, a third God was discovered—the God of Faith, not publically known as the other two constructs but known to individuals as an act of faith.

Emergence maps onto the theological constructs. The reductionist laws of science are statements of the operation of an immanent God. The laws are impenetrable and shrouded in mystery. God's immanence can be studied, and to some this is the scientist's vocation.

The unfolding of the universe of the immanent God into all its novelties and surprises is what we designate as emergence. Each level leads to further complexity. The transition from mystery to complexity would be, in theological terms, the divine spirit. It may be studied in terms of the selection or pruning rules, and thus is also part of the scientist's vocation. All of this is sufficiently new as to be the center of dynamic investigation.

The emergence of the social mind in societies of *Homo sapiens* is the next step. I argue that it is a necessary condition for the divine immanence to become transcendent. Transcendence is only meaningful in a world where the laws of immanence are known, at least partially. To make a very large leap, the emergence of the societal mind resonates with the theologians' concept of "the Son" or "being made in God's image." This argues that the human mind is God's transcendence, and miracles are what humans can do to overcome "the selfish gene" and other such ideas in favor of moral imperatives. The historical Jesus is thus the exemplar of an aspect of divinity that we all share; and a responsibility that we should all undertake. That responsibility is awesome.

REFERENCES

Bateson, Gregory. 1979. *Mind and Nature: A Necessary Unity*. New York: Dutton.

Griffin, Donald R. 1992. *Animal Minds*. Chicago: University of Chicago Press.

Holland, John H. 1998. *Emergence. From Chaos to Order*. Reading, Mass.: Addison-Wesley.

Margenau, Henry. 1977. *The Nature of Physical Reality: A Philosophy of Modern Physics*. Reprint. Woodbridge, Conn.: Ox Bow Press.

Popper, Karl. 1987. "Natural Selection and the Emergence of the Mind." In

Evolutionary Epistemology, Rationality and the Sociology of Knowledge, edited by G. Radnitzky and W. W. Barthley III. La Salle, Ill.: Open Court.

Pullmann, Bernard, ed. 1996. *The Emergence of Complexity*. Proceedings: Plenary Session of the Pontifical Academy of Sciences, 27–31 October 1992. Princeton: Princeton University Press.

Smolin, Lee. 1997. *The Life of the Cosmos*. London: Oxford University Press.

Teilhard de Chardin, Pierre. 1961. *The Phenomenon of Man*. New York: Harper and Row.

TEN

COMPLEXITY, EMERGENCE, AND
DIVINE CREATIVITY

Arthur Peacocke

The Significance of the DNA Structure: Reductionism and Emergence

As my good fortune would have it, when I had completed my doctoral apprenticeship and was for the first time pursuing research entirely of my own devising in my first university post, it was mainly centered on what we now call DNA. In the late 1940s, DNA had been identified as the principal carrier of the genes, but it was still not certain even that it was a large molecule—and although it was known to contain nucleotides linked together in chains of uncertain length, its double-helical structure was unknown. Suffice to say that, after 1952, its discovered structure revolutionized biology and has now become part of general public awareness. What gradually especially impressed itself on me as a physical biochemist participating in this community of discovery is that it is a clue to many important issues in the epistemology and relationships of the sciences—for the first time we were witnessing the existence of a complex macromolecule the *chemical structure* of which had the ability to convey *information*, the genetic instructions to the next generation to be like its parent(s).

In my days as a chemistry student, I had studied the structure of the purine and pyrimidine "bases" which are part of the nucleotide units from which DNA is assembled. All that was pure organic chemistry, with no hint of any particular significance in their internal arrangement of atoms of carbon, nitrogen, phosphorus, and so on. Yet here, in DNA, we had discovered a double string of such units so linked together through the operation of natural processes that each particular DNA macromolecule had the new capacity, when set in the matrix of the particular cytoplasm evolved with it, of being able to convey hereditary information. Now the concept of "information," originating in the mathematical theory of communication, had never been, and could never have been, part of the organic chemistry of nucleotides, even of polynucleotides, that I had learned in my chemistry degree work. Hence in DNA, I realized, we were witnessing a notable example of what many reflecting on the evolutionary process have called *emergence*—the entirely neutral name[1] for that general feature of natural processes wherein complex structures, especially in living organisms, develop distinctively new capabilities and functions at levels of greater complexity. Such emergence is an undoubted, observed feature of the evolutionary process, especially of the biological. As such, it was the goad that stimulated me to wider reflections: first, epistemological, on the relation between the knowledge that different sciences provide; and second, ontological, on the nature of the realities which the sciences putatively claim to disclose—in other words, the issue of reductionism. I am convinced that this has to be clarified in any discussion of complex systems and of their general significance.

There is, as we well know, a polemical edge to all this. For Francis Crick, one of the codiscoverers of the DNA structure, early threw down the gauntlet in these matters by declaring that "the ultimate aim of the modern movement in biology is in fact to explain *all* biology in terms of physics and chemistry" (1966, 10). Such a challenge can be, and has been, mounted at many other interfaces between the sciences other than that between biology and physics/chemistry. We have all witnessed the attempted takeover bids, for example, of psychology by neurophysiology and of anthropology and sociology by biology. The name of the game is "reductionism" or, more colloquially, "nothing-buttery"— "discipline X, usually meaning 'yours,' is really nothing but discipline Y, which happens to be 'mine.'"

After the discovery of the DNA structure, there was an intense discussion between the new "molecular" biologists and "whole-organism" biologists about whether or not Crick's dictum could be accepted. By and large, as evidenced in the book *Studies in the Philosophy of Biology: Reduction and Related Problems* (1974), edited by the geneticists F. J. Ayala and T. Dobzhansky, it was not, for in that book whole-organism biologists insisted on the distinctiveness of the biological concepts they employed and on their irreducibility to purely physicochemical ones. The controversy was, in fact, only one aspect of a debate which had been going on previously among philosophers of science concerning the possible "unity of the sciences," a notion that implied the hegemony of physics in the hierarchy of explanation.

All could, and by and large did, agree on the necessity for *methodological* reduction, that is, the breaking-down of complex systems into their units to begin to understand them and the interrelation of the parts so obtained. That was not the issue. What was at stake was the relation between our knowledge of complex systems and their ontology, between how they are known and how they are conceived actually to be.[2] *Epistemological* reduction occurs when the concepts used to describe and explicate a particular complex system are reducible[3] to, translatable into, concepts applicable to the entities of which the complex is composed. The claim by many biologists, for example, is that many biological concepts are not so reducible.

Ontological reduction is more subtle, being about what complex entities *are*. One form of this simply recognizes, uncontroversially these days (no one claims, for example, to be a vitalist concerning living organisms in the sense of Driesch and Bergson), that everything in the world is constituted of whatever physics says is the basic consitituent of matter, currently quarks. That is to say, most thinkers are, in this respect, monists concerning the ontology of the world and not dualist, vitalist, or supernaturalist. However, to say that diverse entities are all made up of, are "nothing but," the same basic physical constituents is clearly inadequate, for it fails to distinguish and characterize their specific identities and characteristics, that is, their specific ontology. Hence it has come to be widely recognized that this form of basic ontological assertion is inadequate to the complexities of the world, understanding of which can be illuminated by these considerations concerning the varieties of reduction.

It will be enough here to recognize that the natural (and also human) sciences more and more give us a picture of the world as consisting of a complex hierarchy—or, more accurately, hierarchies—a series of levels of organization and matter in which each successive member of the series is a whole constituted of parts preceding it in the series.[4] The wholes are organized systems of parts that are dynamically and spatially interrelated. This feature of the world is now widely recognized to be of significance in coordinating our knowledge of its various levels of complexity—that is, of the sciences which correspond to these levels.[5]

It also corresponds not only to the world in its present condition but also to the way complex systems have evolved in time out of earlier simpler ones.

What is significant about the relation of complex systems to their constituents now is that the concepts needed to describe and understand—as indeed also the methods needed to investigate—each level in the hierarchy of complexity are specific to and distinctive of those levels. It is very often the case (but not always) that the properties, concepts, and explanations used to describe the higher level wholes are not reducible to those used to describe their constituent parts, themselves often also constituted of yet smaller entities. This is an epistemological assertion of a nonreductionist kind, and its precise implications have been much discussed.[6]

When the epistemological nonreducibility of properties, concepts, and explanations applicable to higher levels of complexity is well established, their employment in scientific discourse can often, but not in all cases, lead to a putative and then to an increasingly confident attribution of a distinctive *causal efficacy* to the complex wholes that does not apply to the separated, constituent parts. Now "to be real, new, and irreducible . . . must be to have new, irreducible causal powers" (S. Alexander, as quoted by Kim 1993, 204). If this continues to be the case under a variety of independent procedures[7] and in a variety of contexts, then an *ontological* affirmation becomes possible—namely, that new and distinctive kinds of realities at the higher levels of complexity have *emerged*. This can occur with respect either to moving, synchronically, up the ladder of complexity or, diachronically, through cosmic and biological evolutionary history. This understanding accords with the pragmatic attribution, both in ordinary life and scientific investigation, of the term "reality" to that which we cannot avoid taking account of in our diagnosis of the course of events, in experience or experiments. Real entities have effects and play irreducible roles in adequate explanation of the world.

I shall denote[8] this position as that of *emergentist monism*, rather than as "nonreductive physicalism." For those who adopt the latter label for their view, particularly in their talk of the "physical realization" of the mental in the physical, often seem to me to hold a much less realistic view of higher level properties than I wish to affirm here—and also not to attribute causal powers to that to which higher-level concepts refer.

If we do make such an ontological commitment about the reality of the "emergent" whole of a given total system, the question then arises how one is to explicate the relation between the state of the whole and the behavior of parts of that system at the microlevel. The simple concept of chains of causally related events (A → B → C . . .) in constant conjunction (à la Hume) is inadequate for this purpose. Extending and enriching the notion of causality now becomes necessary because of new insights into the way complex systems, in general, and biological ones, in particular, behave. This subtler understanding of how higher levels influence the lower levels, and vice versa, still allows application in this context of the notion of a kind of "causal" relation between whole and part (of system to constituent)—never ignoring, of course, the "bottom-up" effects of parts on wholes which depend for their properties on the parts being what they are.

The Relation of Wholes and Parts in Complex Systems

A number of related concepts have in recent years been developed to describe the relation of wholes and parts in both synchronic and diachronic systems—that is, respectively, both those in some kind of steady state with stable characteristic emergent features of the whole and those that display an emergence of new features in the course of time.

The term *downward-causation*, or *top-down causation*, was, as far as I can ascertain, first employed by Donald Campbell (1974)[9] to denote the way in which the network of an organism's relationships to its environment and its behavior patterns together determine in the course of time the actual DNA sequences at the molecular level present in an evolved organism—even though, from a "bottom-up" viewpoint of that organism once in existence, a molecular biologist would tend to describe its form and behavior as a consequence of the same DNA sequences. Campbell instances the evolutionary development of efficacious jaws made of suitable proteins in a worker termite. I prefer to use actual complex systems to clarify this suggestion, such as the Bénard phenomenon:[10] at a critical point a fluid heated uniformly from below in a containing vessel ceases to manifest the entirely random "Brownian" motion of its molecules, but displays up and down convective currents in columns of hexagonal cross-section. Moreover, certain autocatalytic reaction systems (e.g., the famous Zhabotinsky reaction and glycolysis in yeast extracts) display spontaneously, often after a time interval from the point when first mixed, rhythmic temporal and spatial patterns the forms of which can even depend on the size of the containing vessel. Many examples are now known also of dissipative systems which, because they are open, a long way from equilibrium, and nonlinear in certain essential relationships between fluxes and forces, can display large-scale patterns in spite of random motions of the units—"order out of chaos," as Prigogine and Stengers (1984) dubbed it.

In these examples, the ordinary physicochemical account of the interactions at the microlevel of description simply cannot account for these phenomena. It is clear that what the parts (molecules and ions, in the Bénard and Zhabotinsky cases) are doing and the patterns they form are what they are *because* of their incorporation into the system-as-a-whole—in fact, these are patterns *within* the systems in question. This is even clearer in the much more complex, and only partly understood, systems of gene switchings on-and-off and their interplay with cell metabolism and specific protein production in the processes of development of biological forms. The parts would not be behaving as observed if they were not parts of that particular system (the "whole"). The state of the system-as-a-whole is affecting (i.e., acting like a cause on) what the parts, the constituents, actually do. Many other examples of this kind could be taken from the literature on, for example, self-organizing and dissipative systems (Gregersen 1998; Peacocke [1983] 1989; Prigogine & Stengers 1984) and also economic and social ones (Prigogine & Stengers 1984).

We do not have available for such systems any account of events in terms of temporal, linear chains of causality as usually conceived (A → B → C → . . .). A wider use of "causality" and "causation" is now needed to include the kind of whole–part relationships, higher- to lower-level, which the sciences have themselves recently been discovering in complex systems, especially the biological and neurological ones. Here the term *whole-part influence* will be used to represent the net effect of all those ways in which the system-as-a-whole, operating from its "higher" level, is a causal factor in what happens to its constituent parts, the "lower" one.

Various interpretations have been deployed by other authors to represent this whole–part relation in different kinds of systems, though not always with causal implications.

Structuring causes. The notion of whole–part influence is germane to one that Niels H. Gregersen has recently employed (1998) in his valuable discussion of autopoietic (self-making, self-organizing) systems—namely, that of *structuring causes*, as developed by F. Dretske (1993) for understanding mental causation. They instance the event(s) that produced the hardware conditions (actual electrical connections in the computer) and the word-processing program (software) as the "structuring causes" of the cursor movement on the screen connected with the computer; whereas the "triggering cause" is, usually, pressure on a key on the keyboard. The two kinds of causes exhibit a different relationship to their effects. A triggering one falls into the familiar (Humean) pattern of constant conjunction. However, a structuring cause is never sufficient to produce the particular effect (the key still has to be pressed); there is no constant relationship between structuring cause and effect. In the case of complex systems, such as those already mentioned, the system-as-a-whole often has the role, I suggest, of a structuring cause in Dretske's sense.

Propensities. The category of "structuring cause" is closely related to that of *propensities*, developed by Karl Popper, who pointed out that "there exist weighted possibilities which are *more than mere possibilities*, but tendencies or propensities to become real" (1990, 12) and that these "propensities in physics are properties *of the whole situation* and sometimes even of the particular way in which a situation changes. And the same holds of the propensities in chemistry, biochemistry and in biology" (17). The effects of random events depend on the context in which they occur. Hence Popper's "propensities" are the effects of Dretske's structuring causes in the case that triggering causes are random in their operation (that is, *genuinely* random, with no "loading of the dice").

For example, the long-term effects of random mutations in the genetic information carrier, DNA, depend on the state of the environment (in the widest sense, so including predators) in which the phenotype comes to exist. This "environment" acts as a structuring cause. Hence a mutation that induces an increase, for example, in the ability of the whole organism to store information about its surroundings might (not necessarily *would*, because of variable exigencies of the environment) lead to the organism having more progeny and so an advantage in natural selection. This is an example of what I have urged (Peacocke 1993) is a propensity in biological evolution. Thus, in this perspective, there are propensities in biological evolution, favored by natural selection, to complexity, self-organization, information processing and storage, and so to consciousness.

Boundary (limiting) conditions. In the discussion of the relations between properties of a system-as-a-whole and the behavior of its constituent parts, some authors refer to the *boundary conditions* that are operating (e.g., Polanyi 1967, 1968).[11] It can be a somewhat misleading term ("limiting condition" would be better), but I will continue to use it only in this wider, Polanyian, sense as referring to the given parameters of the structural complex in which the processes under consideration are occurring.

A more recent, sophisticated development of these ideas has been proffered by Bernd-Olaf Küppers:

> [T]he [living] organism is subservient to the manner in which it is constructed. . . . Its principle of construction represents a boundary condition under which the laws of physics and chemistry become operational in such a way that the organism is reproductively self-sustaining. . . . [T]he phenomenon of emergence as well as that of downward causation can be observed in the living organism and can be coupled to the existence of specific boundary conditions posed in the living matter. (in Russell, Murphy, & Peacocke 1995, 100)

Thus a richer notion of the concept of boundary conditions is operative in systems as complex as living ones. The simpler forms of the idea of "boundary condition," as applied, for example, by Polanyi to machines, are not adequate to express the causal features basic to biological phenomena. Indeed, the "boundary conditions" of a system will have to include not only purely physical factors on a global scale but also complex intersystemic interactions between type-different systems.

There is a sense in which systems-as-a-whole, because of their distinctive configuration, can constrain and influence the behavior of their parts to be otherwise than they would be if isolated from the particular system. Yet the system-as-a-whole would not be describable by the concepts and laws of that level and still have the properties it does have, if the parts (e.g., the ceric and cerous ions in the Zhabotinsky case) were not of the particular kind they are. What is distinctive in the system-as-a-whole is the new kind of interrelations and interactions, spatially and temporally, of the parts.

Supervenience. Another, much-debated, term which has been used in this connection, especially in describing the relation of mental events to neurophysiological ones in the brain, is *supervenience*. This term, which does not usually imply any "whole–part" causative relation, goes back to Donald Davidson's (1980) employment of it in expounding his view of the mind-brain-body relation. The various meanings and scope of the term in this context had been formulated and classified by J. Kim ([1984] 1993) as involving: the supervenient properties' *covariance* with, *dependency* on, and *nonreducibility* to their base properties.

One can ask the question:

> [H]ow are the properties characteristic of entities at a given level related to those that characterize entities of adjacent levels? Given that entities at distinct levels are ordered by the part–whole relation, is it

the case that properties associated with different levels are also ordered
by some distinctive and significant relationship? (Kim 1993, 191)

The attribution of "supervenience" asserts primarily that there is a
necessary covariance between the properties of the higher level and those
of the lower level.

When the term "supervenience" was first introduced it was neutral
with respect to causal relations—of any influence of the supervenient
level on the subvenient one. Later, supervenient causality was even
denied (so Kim). Its appropriateness is obscure for analyzing whole–
part relations, which by their very nature relate, with respect to complex systems, entities that are in some sense the same. For, in the context
of the physical and biological (and, it must also be said, ecological and
social) worlds, the mutual interrelations between whole and parts in
any internally hierarchically organized system often, as I have shown,
appear to involve causal effects of the whole on the parts.

The Mind-brain-body Relation and Personhood

Much of the discussion of the relation of higher to lower levels in hierarchically stratified systems has centred on the mind-brain-body relation, on how mental events are related to neurophysiological ones in
the human-brain-in-the-human-body—in effect the whole question
of human agency and what we mean by it. A hierarchy of levels can be
delineated each of which is the focus of a corresponding scientific study,
from neuroanatomy and neurophysiology to psychology. Those involved in studying "how the brain works" have come to recognize that

> [p]roperties not found in components of a lower level can emerge from
> the organization and interaction of these components at a higher level.
> For example, rhythmic pattern generation in some neural circuits is a
> property of the circuit, not of isolated pacemaker neurons. Higher brain
> functions (e.g., perception, attention) may depend on temporally co-
> herent functional units distributed through different maps and nuclei.
> (Sejnowski, Koch, & Churchland 1988, 1300)

The still intense philosophical discussion of the mind-brain-body relation has been, broadly, concerned with attempting to elucidate the
relation between the "top" level of human mental experience and the
lowest, bodily physical levels. In recent decades it has often involved
considering the applicability and precise definition of some of the terms

used already to relate higher to lower levels in hierarchically stratified systems. The question of what kind of "causation," if any, may be said to be operating from a "top-down," as well as the obvious and generally accepted "bottom-up," direction is still much debated in this context (see, for example, Heil & Mele 1993).

When discussing the general relation of wholes to constituent parts in a hierarchically stratified complex system of stable parts, I used "whole–part influence" (Peacocke 1999b)[12] and other terms and maintained that a nonreductionist view of the predicates, concepts, laws, and so on applicable to the higher level could be coherent. Reality could, I argued, putatively be attributable to that to which these nonreducible, higher-level predicates, concepts, laws, and so on applied, and these new realities, with their distinctive properties, could be properly called "emergent." Mental properties are now widely regarded by philosophers[13] as epistemologically irreducible to their physical ones, indeed as "emergent" from them, but also dependent on them, and similar terms have been used to describe their relation as in the context of nonconscious, complex systems. I have argued (Peacocke 1999b, 229–231) that what happens in these systems at the lower level is the result of the *joint* operation of both higher- and lower-level influences—the higher and lower levels could be said to be jointly sufficient, type-different causes of the lower level events. When the higher–lower relation is that of mind/brain-body, it seems to me that similar considerations should apply.

Up to this point, I have been taking the term "mind," and its cognate "mental," to refer to that which is the emergent reality distinctive especially of human beings. But in many wider contexts, not least that of philosophical theology, a more appropriate term for this emergent reality would be "person," and its cognate "personal," to represent the total psychosomatic, holistic experience of the human being in all its modalities—conscious and unconscious; rational and emotional; spiritual; active and passive; individual and social; and so on. The concept of personhood recognizes that, as Philip Clayton (1998) puts it,

> [w]e have thoughts, wishes and desires that together constitute our character. We express these mental states through our bodies, which are simultaneously our organs of perception and our means of affecting other things and persons in the world. . . . [The massive literature on theories of personhood] clearly points to the indispensability of embodiedness as the precondition for perception and action, moral agency, community and freedom—all aspects that philosophers take as indispensable to human personhood and that theologians have viewed as part of the *imago dei*. (205)

There is therefore a strong case for designating the highest level—the whole, in that unique system that is the human-brain-in-the-human-body-in-social-relations—as that of the "person." Persons are *inter alia* causal agents with respect to their own bodies and to the surrounding world (including other persons). They can, moreover, report on aspects of their internal states concomitant with their actions with varying degrees of accuracy. Hence the exercise of personal *agency* by individuals transpires to be a paradigm case and supreme exemplar of whole–part influence—in this case exerted on their own bodies and on the world of their surroundings (including other persons). I conclude that the details of the relation between cerebral neurological activity and consciousness cannot in principle detract from the causal efficacy of the content of the latter on the former and so on behavior. In other words, "folk psychology" and the real reference of the language of "personhood" are both justified and necessary.

Divine Creativity

We have become accustomed in recent years to hearing of the "epic of evolution" so often that sometimes our ears have become dulled to just how remarkable it is. If something akin to human intelligence had been able to witness the original "hot big bang" some 12 or so billion years ago, would it ever have predicted from the properties of the quarks, the laws of quantum theory and of gravity, and the nature of the four fundamental forces that the process would complexify and self-organize over the aeons in at least one small range of space-time to become persons who can know not only the processes by which they have emerged but also each other and could be creative of truth, beauty, and goodness? It is to the significance of this that we must now turn.

I have been recounting in the foregoing the scientific perspective on a world in which over the course of space-time new realities have emerged by virtue of the inherent properties of basic matter-energy to complexify and self-organize (Gregersen 1998) into systems manifesting new properties and capabilities. These emergent capacities include, we have seen, mental and personal ones and, I would add, spiritual ones—by which I mean the capacity to relate personally to that Ultimate Reality that is the source and ground of all existence. For the very existence of all-that-is, with that inherent creativity to bring persons out of quarks just described, is for me and all theists only explicable by postulating an Ultimate Reality which is: the source and ground of all being and becoming; suprapersonal; suprarational; capable of knowing all that it is logically possible to know and of doing all that it is logically possible to do; and unsurpassedly instantiating the values which human mental and spiritual capacities can discern, if only failingly implement. In English the name for this Reality is "God"—and that usage I will follow from this point.

The question at once arises of how to conceive of the relation of God to all-that-is (the "world"). In more classical terms, how do we conceive of God as *Creator*? The physics of the earlier part of the last century (that is, the twentieth!) showed—as in the famous equation $e = mc^2$—that matter, energy, space, and time are closely related categories in our analysis of the world; so that God must be conceived of as giving existence to, as creating, all time and space as well as matter and energy.[14] So whatever "divine creation" is, it is not about what God can be supposed to have been doing at 4004 BCE or even 12 billion BCE! Divine "creation" concerns the *perennial* relation of God to the world. For we have to conceive now of God giving existence to all entities, structures, and processes "all the time" and to all times as each moment, for us, unfolds. They would not be if God was not. Augustine, of course, perceived this sixteen centuries ago with respect to time when he famously affirmed the impossibility of asking what God was doing "before" creating the world and addressed God thus: "It is therefore true to say that when you [God] had not made anything there was no time, because time itself was of your making" (*Confessions* 11.14, Augustine [c. 400] 1961, 261).

What we now see today, in the light of the whole epic of evolution and our understanding of complex systems, is that the very processes of the world are inherently creative of new realities. We therefore conclude that God is creating all the time in and through the complexifying and self-organizing processes to which God is giving continuous existence in divinely created temporal relations ("time"). God is not a has-been Creator but always and continuously Creator—*semper Creator*, and the world is a *creatio continua*, as traditional theology has sometimes expressed it.

This is far from being a recent concept, for it is implicit in the traditional concept of God's *immanence* in the world. It is noteworthy that, just four years after Darwin published *The Origin of Species*, the Church of England clergyman and novelist Charles Kingsley, in his evolutionary fairy tale *The Water Babies*, depicts Tom, the boy chimney sweep, looking at Mother Earth in puzzlement, for she is apparently *doing* nothing. "Tom to the mother of creation: 'I heard that you were always making new beasts out of old.' 'So people fancy' she said 'I make things make themselves'" (Kingsley [1863] 1930, 248). And Frederick Temple, later the archbishop of Canterbury, affirmed in his 1854 Bampton Lectures that "God did not make the things, we may say, but He made them make themselves" (Temple 1885, 115).

The understanding of cosmic and biological evolution illuminated by new insights into the capacities of complex systems with their self-organizing capabilities and the philosophical framework of an emergentist monism all converge to reinstate the concept of God not only as necessarily *transcendent*—"other" in ultimate Being to be Creator at all—but also as *immanent*: in, with, and under the processes to which God is giving existence. Indeed, these very processes are to be conceived of *as* the activity of God as Creator, and a *theistic* naturalism then becomes imperative. The Christian theological tradition in fact already has imaginative and symbolic resources[15] to enrich this notion:

- creation seen as the self-expression of God the Word/*Logos*
- God's *Wisdom* as imprinted in the fabric of the world, especially in human minds open to "her" (Runge 1999; Deane-Drummond 2000)
- God as the "one in Whom we live and move and have our being" (Paul at Athens in the account of Acts 17:28—a key text in the current reconsideration of "pan*en*theism" as denoting God's relation to the world)
- the tradition's understanding of the sacramental (Temple [1934] 1964, chap. 19; Peacocke 2000);
- in the Eastern Christian tradition, the world as the milieu in which the "uncreated Energies" of God operate (Lossky [1944] 1991)

It is implicit, and is increasingly emphasized recently (Polkinghorne 2000), in the understanding of God's creating that the world is not only dependent on God for its very existence but also that this God-given existence is autonomous in developing its own possibilities by its own inherent, God-endowed capacities and laws. Although the world is in one sense "in" God, as panentheistically understood, yet God is ontologically distinct from it—there is an ontological gap everywhere and at all times between God and the world. Hence creation is a self-limiting activity of God rendering Godself vulnerable, for in it God takes the risk of letting everything be and become itself, and this in human persons, who are free and autonomous, means allowing them to be capable of falsity as well as truth, ugliness as well as beauty, and evil as well as good. God, it appears, literally suffers this to happen for the world to be creative, capable of developing through complexification and self-organization new forms of existence, one of which, *homo sapiens*, is capable of freely chosen, harmonious, personal relations with God's own self.

God is not a magician who overrules by intervening in the creative processes with which God continuously endows and blesses the world—though God is eternally present to it. The future is open, not set in concrete, and does not yet exist even for God to know or determine, but God will, uniquely, be present to all futures and will be able to respond to those personal beings who have evolved to have the capacity freely to respond to God.

The nature of such relationships of persons to God may, like the general scenario of creation outlined in this chapter, also be illuminated by our understanding of the emergence of new realities in complex, especially self-organizing, systems. For in many situations where God is experienced by human persons, we have by intention and according to well-winnowed experience and tradition complexes of interacting personal entities, material things, and historical circumstances that are epistemologically not reducible to concepts applicable to these individual components. Could not new realities—and so new experiences of God for humanity—be seen to "emerge" in such complexes and even to be causally effective?

I am thinking,[16] for example, of the Christian church's Eucharist (Holy Communion, the Mass, "the Lord's Supper"), in which there exists a distinctive complex of interrelations between its constituents. The latter could be identified, *inter alia* (for it is many-layered in the richness of its meanings and symbols), as follows.

1. Individual Christians are motivated by a sense of *obedience* to the ancient, historically well-authenticated command of Jesus, the founder of their faith, at the actual Last Supper to "Do *this* . . ."—that is, to eat the bread and to drink the wine in the same way he did on that occasion and so to identify themselves with his project in the world.

2. Christians of all denominations have been concerned that their communal act is properly *authorized* as being in continuity with that original act of Jesus and its repetition, recorded in the New Testament, in the first community of Christians. Churches have differed about the character of this authorization but not about its importance.

3. The physical "elements," as they are often called, of bread and wine are, of course, part of the matter of the world and so are representative, in this regard, of the created order. So Christians perceive in

these actions, in this context and with the words of Jesus in mind, that a *new significance and valuation of the very stuff of the world* is being expressed in this action.

4. Because it is bread and not wheat, wine and not grapes, which are consecrated, this act has come to be experienced also as a new evaluation of the work of *humanity in cocreating with God in ordinary work*.

5. The broken bread and poured-out wine was explicitly linked by Jesus with his anticipated self-sacrificial offering of himself on the cross, in which his body was broken and blood shed to draw all toward unity of human life with God. Christians in this act consciously acknowledge and identify themselves with Jesus' *self-sacrifice*, thereby offering to reproduce the same self-emptying love for others in their own lives and so to further his purposes of bringing in the Reign of God in the world.

6. They are also aware of the promise of Jesus to be present again in their re-calling and remaking of the historical events of his death and resurrection. This "making-present" (*anamnesis*) of the Jesus who is regarded as now fully in the presence of—and, in some sense, identified with—God is a unique and spiritually powerful feature of this communal act.

7. There is creatively *present* the *God who is transcendent, incarnate,* and *immanent.* Here do we not have an exemplification of the emergence of a new kind of reality, since this complex situation is epistemologically not reducible? For what (if one dare so put it) "emerges" in the eucharistic event *in toto* can only be described in special non-reducible terms such as "Real Presence" and "Sacrifice." A new kind of reality is attributable to the eucharistic event, for in it there is an effect on both the individual and on the community that induces distinctively Christian personhood and society (of "being ever deeper incorporated into this body of love" [Gregersen 2000]). So it is not surprising there is a branch of study called "sacramental theology" to explicate this special reality and human experience and interpretations of it. Since God is present "in, with, and under" this holistic eucharistic event, in it God may properly be regarded as distinctively acting through it on the individual and community.[17]

I have taken this as one example, but I propose that the principle involved in trying to make clear what is special about this particular spiritual situation is broadly applicable[18] to many other experiences of theological concern and interest, both historical and contemporary. For this last reason, in conjunction with the broader exhilarating theistic perspective I have been trying to expound, it seems to me that the new sciences of complexity and of self-organization provide a fruitful release for theology from the oppression of excessively reductionist interpretations of the hierarchy of the sciences and a making-accessible of theological language and concepts to the general exchanges of the intellectual life of our times—a milieu from which it has been woefully and misguidedly excluded for too long.

Would it be too much to suggest that these new, emergentist monist insights into the inbuilt creativity of our world through its complexifying and self-organizing capacities open up a vista of continuity between the physical, the mental, and the spiritual which could, in this new century, break down the parallel barricades mounted in the last, both between the "two cultures" of the sciences and the humanities—and between the experiences of nature and of God, the sciences and religion?

NOTES

1. This term need not (*should* not) be taken to imply the operation of any influences, either external in the form of an "entelechy" or "life force" or internal in the sense of "top-down/whole-part" causative influences. It is, in my usage, a purely descriptive term for the observed phenomenon of the appearance of new capabilities, functions, and so on at greater levels of complexity. It is not intended to have any normative or evaluative connotations.

2. These distinctions were well delineated by F. J. Ayala in his introduction to Ayala and Dobzhansky (1974) and are elaborated in Peacocke ([1986] 1994), chs. 1 and 2).

3. Formal criteria for this had already been developed by Nagel (1952).

4. These are conventionally said to run from the "lower," less complex to the "higher," more complex systems, from parts to wholes so that these wholes themselves constitute parts of more complex entities—rather like a series of Russian dolls. In the complex systems I have in mind here, the parts retain their identity and properties as isolated individual entities. The *internal* relations of such elements are not regarded as affected by their incorporation into the system.

5. See, e.g., Peacocke (1993, 36–43, 214–218, and fig. 3, based on a scheme of Bechtel & Abrahamson 1991, fig. 8.1).

6. For the subtle distinction between "theory" autonomy and "process" autonomy, concepts related to that of ontological reduction, see Peacocke (1999b), (from which some of this text has been drawn), especially the appendix (245–247).

7. W. C. Wimsatt has elaborated criteria of "robustness" for such attributions of reality for emergent properties at the higher levels. These involve noting what is invariant under a variety of independent procedures (summarized in Peacocke [1986]1994, 27–28, from Wimsatt 1981).

8. See Peacocke (1999b) and the discussion of Philip Clayton in the same volume (209–211).

9. For my earlier expositions of mine on the hierarchies of complexity, of the relation of scientific concepts applicable to wholes to those applicable to the constituent parts, and of top-down/downward causation and "whole-part influence," see Peacocke (1993, 39–41, 50–55, 213–218, esp. fig. 3) and Russell, Murphy, & Peacocke (1995, 272–276).

10. For a survey with references see Peacocke ([1983] 1989).

11. In his discussion, and mine in this chapter, the term "boundary condition" is *not* being used, as it often is, to refer *either* to the initial (and in that sense "boundary") conditions of, say, a partial differential equation as applied in theoretical physics *or* to the physical, geometrical boundary of a system.

12. It must be stressed that the "whole–part" relation is *not* regarded here necessarily, or frequently, as a spatial one. "Whole–part" is synonymous with "system-constituent."

13. The "nonreductive physicalist" view of the mental/physical relation of many philosophers has been summarized by Kim (1993), as follows: "1. (*Physical Monism*). All concrete particulars are physical. 2. (*Antireductionism*). Mental properties are not reducible to physical properties. 3. (*The Physical Realization Thesis*). All mental properties are physically realized; that is, whenever an organism, or system, instantiates a mental property M, it has some physical property P such that P realizes M in organisms of its kind. . . . 4. (*Mental Realism*). Mental properties are real properties of objects and events; they are not merely useful aids in making predictions or fictitious manners of speech."

14. For the only dualism acceptable to modern theology and consistent with science is the God–world one, with no fundamental dualities *within* the created world—bearing in mind the nuances and qualifications of the emergentist monism I developed earlier.

15. For a discussion of these ideas, see Peacocke (1999a) and Peacocke (2001), part 3.

16. An interpretation of the Eucharist I originally suggested in Peacocke (1972, especially p. 32) and (with some additions) in Peacocke ([1986] 1994, ch. 9, 124–125). It is entirely congruent with that recently expounded by Gregersen (2000, 180–182).

17. An exemplification of God's nonintervening, but specific, "whole–part" influence on the world, which I have elaborated elsewhere (as in Peacocke 1999b, where references are given)?

18. An approach I have long since adumbrated in my Bampton Lectures of 1978 (Peacocke 1979, app. C, "Reductionism and Religion-and-science: [Theology] the Queen of the Sciences," 367–371).

REFERENCES

Augustine. [c. 400] 1961. *Confessions*. Translated by R. S. Pine-Coffins. Penguin Classics edition. Harmondsworth: Penguin.

Ayala, F. J., and T. Dobzhansky, eds. 1974. *Studies in the Philosophy of Biology: Reduction and Related Problems*. London: Macmillan.

Bechtel,W., and A. Abrahamson. 1991. *Connectionism and the Mind*. Oxford: Blackwell.

Campbell, D. T. 1974. " 'Downward Causation' in Hierachically Organised Systems." In *Studies in the Philosophy of Biology: Reduction and Related Problems*, edited by F. J. Ayala and T. Dobhzhansky. London: Macmillan.

Clayton, Philip. 1998. "The Case for Christian Panentheism." *Dialog* 37, 3: 201–208.

Crick, Francis H. C. 1966. *Of Molecules and Man*. Seattle: University of Washington Press.

Davidson, D. 1980. "Mental Events." In *Essays on Actions and Events*. Oxford: Clarendon Press.

Deane-Drummond, Celia. 2000. *Creation through Wisdom: Theology and the New Biology*. Edinburg: T. and T. Clark.

Dretske, Fred. 1993. "Mental Events as Structuring Causes of Behavior." In *Mental Causation*, edited by J. Heil and A. Mele. Oxford: Clarendon Press.

Gregersen, Niels Henrik. 1998. "The Idea of Creation and the Theory of Autopoietic Processes." *Zygon* 33: 333–367.

Gregersen, Niels Henrik. 2000. "God's Public Traffic: Holist versus Physicalist Supervenience." In *The Human Person and Theology*, edited by N. H. Gregersen, W. B. Drees, and U. Görman. Edinburg: T. and T. Clark/Grand Rapids: Eerdsman.

Heil, J., and A. Mele, eds. 1993. *Mental Causation*. Oxford: Clarendon Press.

Kim, J. 1993. "Non-Reductivism and Mental Causation." In *Mental Causation*, edited by J. Heil and A. Mele. Oxford: Clarendon Press.

Kim, J. [1984]1993. "Epiphenomal and Supervenient Causation." *Midwest Studies in Philosophy 9* (1984): 257–270. Reprinted in *Supervenience and Mind: Selected Philosophical Essays*. Cambridge: Cambridge University Press.

Kingsley, Charles. [1863] 1930. *The Water Babies*. London: Hodder and Stoughton.

Lossky, Vladimir. [French ed. 1944] 1991. *The Mystical Theology of the Eastern Church*. Cambridge: James Clark.

Nagel, Ernest. 1952. "Wholes, Suns and Organic Unities." *Philosophical Studies* 3: 17–32.

Peacocke, Arthur. 1972. "Matter in the Theological and Scientific Perspectives." In *Thinking about the Eucharist*, a collection of essays by members of the Doctrine Commission of the Church of England. London: SCM Press.

Peacocke, Arthur. 1979. *Creation and the World of Science*. Oxford: Clarendon Press.

Peacocke, Arthur. [1983] 1989. *The Physical Chemistry of Biological Organization*. Reprinted with additions. Oxford: Clarendon Press.

Peacocke, Arthur. [1986] 1994. *God and the New Biology*. London: Dent 1986; reprint, Gloucester, Mass.: Peter Smith.

Peacocke, Arthur. 1993. *Theology for a Scientific Age: Being and Becoming—Natural, Divine and Human*. 2nd enl. ed. Minneapolis: Fortress Press/London: SCM Press.

Peacocke, Arthur. 1999a. "Biology and a Theology of Evolution." *Zygon* 34: 695–712.

Peacocke, Arthur. 1999b. "The Sound of Sheer Silence: How Does God Communicate with Humanity?" In *Neuroscience and the Person: Scientific Perspective on Divine Action*, edited by R. J. Russell et al. Vatican City: Vatican Observatory.

Peacocke, Arthur. 2000. "Nature as Sacrament." *Third Millennium* 2: 16–31.

Peacocke, Arthur. 2001. *Paths from Science towards God: The End of All Our Exploring*. Oxford: Oneworld.

Polanyi, M. 1967. "Life Transcending Physics and Chemistry," *Chemical and Engineering News*, August 21.

Polanyi, M. 1968. "Life's Irreducible Structure." *Science* 160: 1308–1312.

Polkinghorne, John C., ed. 2000. *The Work of Love: Creation as Kenosis*. Grand Rapids, Mich.: Eerdmans/London: SPCK.

Popper, Karl. 1990. *A World of Propensities*. Bristol, England: Thoemmes.

Prigogine, I., and I. Stengers. 1984. *Order out of Chaos*. London: Heinemann.

Runge, Sharon H. 1999. *Wisdom's Friends*. Louisville, Ky.: Westminster John Knox Press.

Russell, R. J., N. Murphy, and A. N. Peacocke, eds. 1995. *Chaos and Complexity: Scientific Perspectives on Divine Action*. Vatican City State: Vatican Observatory/Berkeley: CTNS.

Sejnowski, T. J., C. Koch, and P. Churchland. 1988. "Computational Neuroscience." *Science* 241: 1299–1306.

Temple, Frederick. 1885. *The Relations between Religion and Science*. London: Macmillan.

Temple, William. [1934] 1964. *Nature, Man and God*. London: Macmillan.

Wimsatt, W. C. 1981. "Robustness, Reliability and Multiple-determination in Science." In *Knowing and Validating in the Social Sciences: A Tribute to Donald T. Campbell*, edited by M. Brewer and B. Collins. San Francisco: Jossey-Bass.

ELEVEN

FROM ANTHROPIC DESIGN TO
SELF-ORGANIZED COMPLEXITY

Niels Henrik Gregersen

God is in the richness of the phenomena, not in
the details of science.

Freeman J. Dyson

Around 1980 a fascinating new species of scientific exploration emerged: the computer-aided studies of complex, self-organizing systems.[1] *Complex* systems consist of a huge number of variegated and constantly interacting elements. Their elements and agents are too many and too different to be appropriately analyzed in a bottom-up way, and unexpected properties emerge as the concerted result of the interactions within the systems. Think of the circularity between RNA and DNA in cells, the interaction of ants in their communities, neurons in brains, or the myriads of interactions in modern city life. Even though complex systems are highly different in nature and scope, the general lesson is that more comes out of less. While the material constituents remain the same, the organization of matter grows exponentially, and does so by self-organization.

Now *complexity studies* attempt to understand the principles guiding such complex systems in order to explain how it can be that ordered structures organize themselves without any controlling consciousness to oversee the process. No master ant in the hill, no designer neuron in the brain, no mayor capable of knowing and controlling the untamable life of a metropolis. Nonetheless, ordered structures are built up, sustained, and further developed in a process driven by the local behavior of the individual agents within the system. In fact, local twists and turns

206

often bring about far-reaching consequences for the system as a whole. Complexity studies want to understand the rules for the propagation of order in "real-world" systems, natural as well as social. The highway of understanding, however, goes through computer modeling.

So far complexity research has some affinities to *chaos theory*. Both fields are spinoffs of the new revolutionary use of computers in science. Both fields deal with nonlinear processes in which small and simple inputs can lead to large and complex outputs. There are, however, important differences as well. Whereas the distinctive trajectories of chaotic systems are highly contingent upon the exact values of the initial conditions, self-organized complex systems are more *robust*, that is, they can take off from a broad variety of initial conditions. Moreover, ordered complexity only emerges in rather limited domains of the chaotic systems governed by the Lyapunov exponent. Since the building-up and propagation of complex orders nonetheless appear almost ubiquitously in nature, complexity seems to be generated by more general principles. In this perspective, mathematical chaos theory appears to be a subsection under complexity theory. Even if chaotic states (in the nontechnical sense of the term) appear also within complexity, the specific mathematics of chaos cannot explain the emergence of complex systems as a whole (Bak 1997, 29–31).

The Hypothesis: A World Designed for Self-Organization

In the following pages I want to explore the religious significance of complexity studies. Before doing so, let me review the theological challenge involved in the notion of self-organizational processes: Their robustness might seem to make any reference to an external creator obsolete. In fact, the title of this essay could be read in terms of a replacement hypothesis: "from design to self-organization"; it is meant, however, to indicate a direction or flow from the "initial" design of the laws and basic properties of physics to self-organizational processes. Thus, I am going to argue that on the basis of a prior assumption of God's benevolence and generosity, we should naturally be inclined to think of self-organization as the apex of divine purpose. Making room for otherness is logically implied by the idea of creation; making the creatures make themselves can be seen as a further emphasis on the autonomy generously bestowed on creatures. Thus we might be designed for self-organization. God's design of the *world as whole* favors the emergence of autonomous processes in the *particular course* of evolution, a course at once constrained and propagated by a built-in propensity toward complexification. Rather than seeing self-organization as a threat to religion, we should see God as continuously creating the world by constituting and supporting self-organizing processes (Gregersen 1998, 1999).

The ideas of design and self-organized complexity, however, are only compatible on two conditions that specify their respective domains of applicability: (1) The notion of divine design relates to the constitution of the world of creation *as a whole* and to the coordination of the basic laws of nature but not to the *details* emerging within the framework of the world. Accordingly, even if the basic laws of physics were to be unified in a grand unified theory, they would not give us a sufficient scientific explanation for *particular features* within cosmic evolution, such as the informational structure of the DNA, the upright gait of human beings, or the AIDS epidemic (see Barrow 1991, 40). As argued by Stuart Kauffman in this volume, if the world is tuned for the emergence of autonomous agents, we cannot prestate the "configuration space" of the biosphere. (2) Conversely, the concept of self-organization should not be elevated into a metaphysical principle that is able to explain all-that-exists. There is no observational basis for claiming that either selection processes or principles of self-organization are responsible for laws of physics such as gravity, quantum mechanics, cosmological constants, and so on. In fact, standard science offers good evidence for believing that some laws of physics are basic and that it is these laws that make our planet (and perhaps other planets) habitable. Neither selection nor self-organization start from scratch; rather, they presupposes a sufficient flow of energy and are channeled within an already existing order.

I thus believe that there are good reasons for steering a middle course between two parties the discussion on religion and self-organization: those who believe that self-organization simply replaces the idea of a divine designer and those who believe that unless the belief in God is cashed out in some supernatural explanation of particular affairs of nature (in contrast to naturalistic explanations), God is pushed out from the world of nature. Thus, the two positions I want to argue against are on the one hand the proponents of a replacement hypothesis and on the other the proponents of the anti-Darwinian hypothesis of "intelligent design." Both parties seem to me to make a common fallacy of misplaced concreteness: If God is not to be traced in the details of scientific exploration of nature, God cannot be present in the tissues and texture of the world.

Self-Organization Does Not Mean Self-Creation

According to the replacement thesis, William Blake's famous picture of the Geometer God ("Europe," 1794) has become meaningless in a self-organizing world. Nature herself does the job that God was once assigned to do. This metaphysical position has recently been uncompromisingly put by the cosmologist Lee Smolin as follows.

> What ties together general relativity, quantum theory, natural selection, and the new sciences of complex and self-organized systems is that in different ways they describe a world that is whole unto itself, without any need of an external intelligence to serve as its inventor, organizer, or external observer. (Smolin 1997, 194; see Smolin 2000, 84–85)

This is clearly a highly metaphysical interpretation of science, since one could hardly say that any such conclusion is implied by any of the aforementioned sciences. It is true that the physical sciences are silent about God, but this silence is a methodological requirement that offers no evidence for or against the reality of the divine. Smolin's judgment rests on his particular hypothesis that not only are systems self-organizing but so are also the physical laws of nature. According to Smolin, all laws of nature have been pruned in a grand cosmological selection process beyond our reach. Consequently, basic laws of nature do not exist. Interesting as this philosophical hypothesis is, it is definitely not implied by either quantum theory, relativity, or evolutionary biology— or by complexity studies. Even strong critics of religion such as Steven Weinberg admit that "if we were to see the hand of the designer anywhere, it would be in the fundamental principles, the final laws of nature, the book of rules that govern all natural phenomena" (1999, 10). Smolin's argument against design rests on the metaphysical assumption that there exists no basic framework of laws to be explained. Because the *explanandum* has disappeared, we should also take leave of the *explanans*. Unless one is prepared to follow this speculative proposal to its very end, there is no inherent conflict between the traditional doctrine of creation and the principles of self-organizing systems.

Naturalistic versus Antinaturalistic Design Theories

As is evident from the last 30 years of theistic interpretation of the anthropic principle, the classic idea of God as designer of the universal laws is far from outmoded, even though the idea of design (as I will show) is not without alternatives. More dubious is a more recent version of design arguments propelled by the so-called intelligent design movement. According to this group of scholars, a divine design can be inferred from particular features of nature insofar as these resist a full explanation in naturalistic terms. The intelligent design theorists remind us that there exist important gaps in the scientific knowledge of nature, especially about the origins of life. As a matter of fact, no one has yet come up with a satisfactory theory about how the interaction between DNA and RNA came about in the first place.[2] They also rightly point out that the sequential order of the DNA molecules (the "specified information") cannot be derived from the chemical affinities of the constituent building blocks themselves; if this were the case, evolution would take place in a chemical straitjacket that would not allow for much variation (Meyer 1998, 132–133). I believe, however, that the importance of the many existing subexplanations, for example, on protometabolism (see de Duve 1998), is either ignored or belittled. Intelligent design theorists seem to be playing an all-or-nothing game. Since we have in our hand *no full* explanation of biogenesis in terms of biochemistry, we have *no* explanations *at all*. But, as has rightly been pointed out by critics of the intelligent design theories, "complexity" is a redundant phenomenon of ordinary chemistry, and the "specificity" of information is what one would expect of a DNA profile that has been carved out through a long history of variation and selection (Shanks & Joplin 1999). Unsatisfied by the partial explanations of current science, the intelligent design theorists offer a wholly other type of explanation: intelligent design. Just as we are used to detecting intelligent agency in everyday life as well as in the social sciences, so we can infer a divine mind from the subtlety of "specified information" in the DNA world. What could not be explained naturalistically is economically explained by design. So the argument goes.

To me this is an example of an impatient science that changes the level of explanation from natural causes to philosophy of mind when (wherever?) there are gaps in the explanatious of current science. My main objection, however, is of a theological nature. If one wants to speaks of divine design, God's purposes may well be compatible with both law and chance. In fact, God's constituting of the basic laws of nature makes up a core tenet in the inherited notion of design. But, as often argued by Arthur Peacocke (2001, 75–78), it is the intricate interplay between laws (which guarantee the overall order) and chance (which introduces novelty into the world) that drives evolution forward. In fact, the open-endedness of evolutionary processes (within a given phase space) is highly congruent with the idea of a benevolent God. Who, by analogy, are the more loving parents: those who specifically instruct their children to become, say, lawyers, or those who let their children explore their individual possibilities within a well-proportioned balance of safe background conditions and an influx of time and circumstance?

Perhaps the most distinctive move of the intelligent design movement lies in the presupposition that divine design can be best (or only) detected in the absence of naturalistic explanations. William Dembski's "explanatory filter" of design, via the exclusion of first law and then chance, seems to rely on a competitive view of God and nature (see Dembski 1999; compare Dembski 1998, 55–66).

Hereafter I intend to show that this presupposition is not in accord with the general thrust of a religious way of perceiving nature. The logic of both the Jewish Bible on creation and the New Testament texts on the Kingdom of God presuppose that the more creative nature is, the more benevolent and the more beautiful is the grandeur of God's creativity. Similar views can be found in other traditions as well. This internal religious perspective can, I believe, be reformulated in the context of a philosophical theology. Bringing the anthropic principle in communication with the theory of self-organized complexity thus may give theology rich resources for redescribing, in religious terms, a cosmos that has already been described and (at least partially) explained by the sciences. There is no need to turn to the oppositional thought pattern of God versus nature. By contrast, my hypothesis is that God's splendor is enhanced by the capacities unveiled by the evolutionary history of our cosmos.

The Internal Religious Perspective

Let us take a closer look at the logic that persuaded some to think that the principles of self-organization have devastating implications for religion. Simply put, the argument was as follows: "If nature makes itself, God does not make nature."

God's Transcendence and Immanence in an Inventive Universe

In 1975 the physicist K. G. Denbigh, in his widely read book *An Inventive Universe*, made the claim that the denial of nature's self-generative powers is a logical consequence of the idea of divine transcendence. According to Denbigh, God's transcendence means that God has created, and still rules, the world "as if from the outside"; this religious view is then said to lead on to the idea that the world's material constituents have no creative powers of their own; nothing essentially new could ever be produced by matter (Denbigh 1975, 11; see 149).

It should be conceded that, on a theistic view, nothing happens in separation from God; but Denbigh's alternative—God or nature—misses the religious point of view. In fact, according to standard theism, God is believed to at once transcendent as well as immanent. Here is just one example from the New Testament.

> The God who made the world and everything in it, he who is Lord of heaven and earth, does not live in shrines made by human hands, nor is he served by human hands, as though he needed anything [transcendence!]. . . . [I]ndeed he is not far from each one of us [immanence]. For "In him we live and move and have our being," as even some of your poets have said, "For we too are his offspring." (Acts 17:24–28)

Thus God is perceived to be infinitely beyond any empirical event, yet God is also qualitatively present in the world without losing God's self-identity. Conversely, the world also may be said to exist in God (a view often termed panentheism).

Both in the scientific and philosophical literature on self-organization we nonetheless often find views like Denbigh's. Lee Smolin has been mentioned already; another example is the philosopher-engineer Paul Cilliers, who in passing claims that self-organization implies that "nothing 'extra,' no external telos or designer is required to 'cause' the complex behavior of a system" (Cilliers 1998, 143). This statement holds true within an empirical context, but the sentence wrongly suggests that God is denied by way of logical implication: "If self-organization, then no design."

However, the alternative "divine design versus natural self-organization" only appears if one fails to distinguish between the theory level of science and the worldview assumptions that may (or may not) be associated with a given domain of science. The search for scientific explanations is always pursued under the condition of a *methodological naturalism*. Therefore, God cannot and should not be found in the epistemic gaps left over by the limits of current scientific explanation. Such methodological naturalism may also include the stronger ontological stance that David Ray Griffin has dubbed a *minimal naturalism*, namely, the assumption that "the world's most fundamental causal principles are never interrupted" (Griffin 2000, 44). But this by no means rules out the view that the observed processes of self-organization are placed in the framework of cosmic conditions that could well be in need of a metaphysical or religious explanation.

God: The Source of Self-Organization

Indeed, the idea of self-organization is far from foreign to the great world religions. In Hindu scriptures, such as the *Rig-Veda* (10.129), we find the notion that in the beginning a golden germ of fire sprang up within the water. This could well be interpreted as an expression of creation as rooted in emergent processes. Later in the Upanishads, we find the idea of "the egg of Brahman," which implies that the world starts off as an undivided whole that is then subsequently divided and complexified (*Aitareia Brahmana* 2.17.8; Gombridge 1975, 114–118).

But also in the Jewish and Christian tradition the notion of spontaneous generation is present. As pointed out by Rabbi Louis Jacobs, the Hebrew word for creating, *bara*, literally means the "cutting out" of existing material (Jacobs 1975, 71). In the book of Genesis, God is portrayed as the one who grants the creatures a power to emerge and reproduce. God creates by letting the earth bring forth vegetation with self-sustaining capacities: "Then God said, 'let the earth put forth vegetation: plants yielding seed, and fruit trees of every kind on earth that bear fruit with the seed in it'" (Genesis 1.11). Biblical scholars believe that we here find traces of a Mother Earth mythology that has been inscribed into the theology of creation. Be that as it may, it is evident that God is depicted as the one who elicits the created powers in order to let them grow, increase, and be fruitful. God's generosity is highlighted; accordingly, the motif of God's blessing of the creaturely powers is emphasized.

If we turn to the New Testament, we also find that the Kingdom of God (i.e., God's active and qualitative presence in the world) is likened to the self-productive capacities of nature. In the teaching of Jesus, we find the following parable.

> The Kingdom of God is as if someone would scatter seed on the ground, and would sleep and rise night and day, and the seed would sprout and grow, he does not know how. The earth *produces of itself* [Greek: *automatike*], first the stalk, then the head, then the full grain in the head. (Mark 4:26–28)

As is evident, neither the Jewish nor the Christian scriptures conceive of God as being in competition with nature's capacities; rather, God is the facilitator of self-generative processes manifest in nature. Against widespread opinion, there is not a hint of contradiction between God's creativity and the creature's self-productivity. Rather, we seem to be facing a two-phase relation between God and world: First, God unilaterally creates the world and its capacities; second, creatures partake in a bilateral flow between divine and natural powers (see Welker 1991, 56–71).

Combining Anthropic Principle and Self-Organized Complexity

In what follows I wish to propose a differentiated notion of God's relationships to an evolving world. By "differentiated" I mean that theology should be able to employ different thought models in relation to different problems. The idea of design, as we have seen, may be illuminating with respect to the fundamental structure of reality; but the inherited notion of design seems to be misplaced in relation to self-organizational processes. My proposal is therefore that theology should move beyond a stereotypical use of its concepts and theoretical models. In some contexts, the notion of design may be appropriate, in other contexts it may be unpersuasive.

Limiting the Metaphor of Design in Theology

The idea of a divine master plan is one standard model of design, derived from the field of human prudence. As phrased by Thomas Aquinas, "it is necessary that the rational order of things [*ratio ordinis*] toward their end preexists in the divine mind." The preconceived divine plan is then presumed to be carried out (*executio*) in the course of history, though in such a manner that God normally uses natural means (*executrices*) to accomplish the divine purpose (*Summa Theologica* 1.22.3). This teleological-instrumental model of divine activity may have its uses, but since it is framed within a deterministic thought model, it is often generalized and used without acknowledging its internal limitations. Accordingly, the design idea tends to have a very low informational value.[3]

Of course, one can hold the position that the basic physical laws are divine instruments, and *so* are the mathematical orders of complexity, and *so* are the Darwinian principles of selection and chance, and *so* are the workings of the human brain. This strategy, however, seems feeble from a theological perspective; moreover, it is incapable of differentiating between questions pertaining to cosmology (concerning the basic laws of physics) and questions pertaining to self-organizing systems (concerning the regularities of complexity). Explaining the framework of the world as such does not always explain the particular features emerging *within* that framework.

In what follows I will therefore argue for a more limited use of the concept of design in theology. More precisely, I want to bring a theistic interpretation of the anthropic principle into conversation with the paradigm of self-organized complexity. The fact of anthropic coincidences fits well with religious expectations of how a generous God would set up a world. However, the robustness of self-organizing systems seems to make a design hypothesis superfluous. I am nonetheless going to argue that the hypothesis of a divine design of all-that-exists is corroborated by the fecundity and beauty that result from the working of self-organizational principles. From a theological perspective, the effectiveness of self-organization may be seen as exemplifying a principle of grace written into the structure of nature. Self-organizational processes are always risky and fragile, but cooperation and the building-up of higher-level structures do pay off in the long run. Without claiming to "explain" the causal routes of self-organization in religious terms, the fertility of self-organizational processes is seen as the blessing of nature by a generous God (Gregersen 2001a).

The Strong Version of the Anthropic Principle

According to the anthropic principle, the nature of the physical universe is highly constrained by the fact that we are here to observe it. There is a link between the basic laws and constants of our physical cosmos and our being as human observers (therefore the term "anthropic" cosmological principle). The fact of intelligent life tells a story about the kind of universe we are inhabiting. The question, then, is how strong or weak the connection is and whether one can legitimately infer any theological conclusions from the many anthropic coincidences. According to the strong anthropic principle (SAP):

> The universe must have those properties which allow [intelligent] life to develop within it at some stage of its history.[4]

On this view the constants and laws of nature are *necessarily* as life-supporting as they are because we are here to observe the universe. To paraphrase Descartes, "I think; therefore, the world is as it is." On this view, the interrelation between the cosmos and intelligent life (such as ourselves) imposes restrictions not only on our location in the universe (we could, for example, not have arrived five billion years earlier) but also on the fundamental parameters of the universe. This point was already made by the cosmologist Brandon Carter when he first, in 1974, introduced the idea of the anthropic principle ([1974] 1998, 134–135).

The large number of cosmic coincidences are open to a religious interpretation: our particular universe may be fine-tuned by God for the purpose of generating intelligent beings. Obviously, this is a metaphysical and not a physical explanation, but it is one that explains the particular features of the anthropic–cosmological coincidences that cannot easily be explained within a physical framework; after all, it is the very framework of physics that calls for an explanation.

This immediate theistic interpretation, however, is not without alternatives. The most widespread is the so-called many world interpretation (MWI), which holds the following position:

> An ensemble of other possible universes is necessary to explain the fact that our particular universe is as life-supporting as it is.

On this interpretation, we are simply the lucky inhabitants in a universe (or in one of many universes) that seems *as if* geared to the emergence of life. The existence of many universes can then be imagined either as standing in a temporal sequence (oscillating universes replacing one another) or as simultaneous universes placed in different regions of a larger universe of universes. So far, the MWI appears to be a purely metaphysical hypothesis, which, compared with the design hypothesis, has the disadvantage of being construed purely ad hoc, for the purpose of explaining the "problem" of the large number of cosmic coincidences without explaining anything else. On this score, the traditional designer hypothesis may be judged not only as more economical but also as capable of explaining far more features than the MWI (e.g., the comprehensibility of the world, the beauty of the universe, the overall progress of evolution, the urge to love another, the independent attestation of religious experience, etc.).

However, quantum versions of the MWI have been proposed that have a stronger linkage to scientific theory building. Already in the 1950s Hugh Everett argued that quantum mechanics supplies a mechanism for generating separate worlds. His proposal was that wave-functions never truly collapse (as is presupposed in the Copenhagen interpretation). Rather, at each quantum time the world splits into branches that thereafter hardly interact. With one notable exception, though. According to Everett, the existence of other universes explains the otherwise unexplained effects of quantum wave interference in the famous double

split experiment (Barrow & Tipler [1986] 1996, 458–509). More recently, David Deutsch has revived Everett's ontological interpretation
of quantum theory. Deutsch explains the observations of quantum wave
interference as an interference between the single photon (that has been
detected by the measuring apparatus as passing through one of the two
slits) with other undetectable photons in an infinity of adjacent universes. These universes are claimed to be real, since only real things
can have real effects: the wave interference. In this manner, Deutsch
infers a whole worldview on the basis of the double split experiment.
Our so-called universe is in reality a "multiverse," that is, a multilay-
ered structure of quantum time instants. Each time (each snapshot of
observation) constitutes a whole universe of its own, without any
overarching framework of time or observational perspective. What is a
shadowy universe to one quantum event is the real universe to another,
but all possible states of quantum processes are fully real.

> We exist in multiple versions, in universes called "moments." Each
> version of us is not directly aware of the others, but has evidence of
> their existence because physical laws link the contents of the different
> universes. It is tempting to suppose that the moment of which we are
> aware is the only real one, or is at least a little more real than the others.
> But that is just solipsism. All moments are physically real. The whole
> of the multiverse is physically real. Nothing else is. (Deutsch [1997]
> 1998, 287)[5]

It is of course highly adventurous to draw such far-reaching ontological messages on the basis of the features of wave interference. But
this is not my concern here. The question is what the theological options are for dealing with the many world assumption. One option is
to understand the idea of God as an explanatory rival to the MWI. This
strategy, however, would only be legitimate if God's vision for the world
were confined to letting into being just one single universe, for which
God has a very definite "plan." But is this the only theological option?
Certainly, this was the notion of design that has been presupposed in
the influential strands of religious determinism (for example, in Islam
and in some forms of Christian tradition). But a deterministic design
theory also can be coupled with the MWI, as in the Stoic doctrine of
oscillating universes in the context of a deterministic cosmology. But
indeterminism also is a theological truth candidate, and indeed a more
flexible one. If we understand God as the creator of creativity (Gregersen
2001), God's intention is to let the world flourish according to its own,
God-given possibilities. In Jewish tradition, we thus find the idea that

God blesses the world already created by God. In the Neoplatonic tradition we find the idea of the "principle of plenitude," which has influenced Christian philosophers from Augustine to Leibniz (Lovejoy [1936] 1964). This understanding of a divine design as a creative and self-giving design is highly compatible with the MWI. Indeed, the theological notion of God's superabundance would be open to many forms of divine creativity, in this universe as well as in other universes. Even though this remains a thought experiment (since we cannot know about other universes), we find in the Narnia books of C. S. Lewis a modern awareness of the possibility of multiple parallel universes, in the context of an orthodox Christian thinker. As we will see, the idea of design can be placed at higher logical levels than the notion of one particular design for one particular world.

Michael Denton's Naturalistic Version of the Strong Anthropic Principle (SAP)

I shall nonetheless stay within the boundaries of the observable universe and here focus on the degree of coordination between the laws of physical chemistry and biological life. In *Nature's Destiny. How the Laws of Biology Reveal Purpose in the Universe*, the biochemist Michael J. Denton has pointed to the long chain of anthropic coincidences, from the basic laws of physics to the life-supporting contribution of sunlight and the hydrosphere, and to the aptness of water and especially carbon as bearers of intelligent life. None of these properties are created by Darwinian evolution, since they precede biological life. Denton's observations are by no means new, taken individually, but he should be credited with having added further weight to the hypothesis that life is the property of a uniquely concerted universe. He points to the fact, for instance, that silicon would not work as a replacement for carbon, since "it falls far short of carbon in the diversity and complexity of its compounds." In particular, he underlines that carbon's aptness for life "is maximal in the same temperature range that water is fluid," which is not the case for silicon (Denton 1998, 101; see 109–116).[6]

Against this background, Denton argues that the cosmos is a "uniquely prefabricated whole with life as it is on earth as its end and purpose" (Denton 1998, 367). This is the SAP; for Denton's thesis is (1) that intelligent humanlike life will *necessarily* arise given the way the world is construed and (2) that the physical-chemical structure of the universe has been *uniquely designed* for that unique purpose.

Denton is aware that his position cannot be empirically proven; after all, a necessity does not follow from even a very long chain of contingent coincidences. But he underlines that his hypothesis could be falsified, for example, by the evidence of extraterrestrial life on noncarbon templates. Furthermore, Denton makes clear that his teleological argument (unlike the intelligent design theory) is developed on fully naturalistic foundations, namely, that our cosmos is "a seamless unity which can be comprehended ultimately in its entirety by human reason and in which all phenomena, including life and evolution, are ultimately explicable in terms of natural processes" (Denton 1998, xviii). Denton thus escapes both the Scylla of a principled antiteleology and the Charybdis of a design theory that erases naturalistic explanations. Denton carefully distinguishes the causal explanations of science from the semantic explanation of a natural theology.[7] Therefore, one could subscribe to Denton's long list of cosmic coincidences without necessarily explaining them in terms of divine purpose. The fact that intelligent life *must* happen, given the physics we have, does not prove that we *must* have the physics we have by virtue of God's benevolent design. The category of design certainly has an illuminating power, but there is no cogent inference to be made from the existence of a vast number of lucky coincidences to the existence of a designing God.

Another strength of Denton's position is that he avoids the view that Barrow and Tipler named the final anthropic principle (FAP):

> Intelligent information-processing must come into existence in the Universe, and, once it comes into existence, it will never die out. (Barrow & Tipler [1986] 1996, 23)

The latter part of the sentence simply overstates what we know about the cosmic conditions. An eternal future of life processes cannot be deduced from the many happy coincidences of past cosmic history.

George Ellis's Theological Interpretation of the Weak Anthropic Principle (WAP)

The weak anthropic principle (WAP) has more followers; according to this version, the universe can only be understood as *actually* providing the right physical conditions for intelligent life. Barrow and Tipler put the principle as follows.

> The observed values of all physical and cosmological quantities are not equally probable but they take on values restricted by the requirements that there exist sites where carbon-based life can evolve and by the

requirement that the Universe be old enough for it to have already done so. (Barrow & Tipler [1986] 1996, 16)

This weak version has sometimes been called a truism. Barrow and Tipler thus described the WAP as a mere restatement of a well-established principle of science, namely, to take into account the limitations of one's measuring apparatus when interpreting one's observations. As pointed out by the South African cosmologist George Ellis, however, the weak version can be seen as a meaningful research program for understanding the road from physics to life by asking "[h]ow much variation in laws and initial conditions can there be, and still allow intelligent life to exist, including the unique features of the mind such as consciousness and self-consciousness?" (Ellis 1993, 377)

As a matter of fact, most of the basic observations of the cosmological coincidences that favor life are not questioned in the scientific literature; it is the interpretation of the facts (especially the SAP and FAP versions of the anthropic principle) that have been under attack. The existence of consciousness does indeed impose extremely narrow conditions on the basic laws of the universe (such as gravity, weak and strong nuclear force) and on its initial conditions (such as size), *if* life and intelligence is to evolve. These coincidences include the following features:[8]

- To avoid a nearly immediate recollapse of the cosmos, the *expansion rate* at early instants needed to have been extremely fine-tuned (perhaps to one part in 10^{55}).
- Had the *nuclear weak force* been a little stronger, the big bang would have burned all hydrogen to helium; no stable stars could then have evolved. Had the force been weaker, the neutrons formed at early times would not have decayed into protons.
- For carbon to be formed within stars, the *nuclear strong force* could vary only up to 1–2 percent in each direction.
- Had *electromagnetism* been appreciably stronger, stellar luminescence would fall, and all stars would be red stars, probably too cold to support the emergence of life and unable to explode as supernovae. Had electromagnetism been slightly weaker, the main sequence stars would be too hot and too short-lived to sustain life.
- *Gravity* also needs to be fine-tuned and correlated with electromagnetic forces in order to create long-lived stars that do not burn up too early.

In a second step, these extremely delicate balances of the life-giving physical condition may then be taken as a warrant for the widespread religious belief that a divine designer is responsible for creating and maintaining the extraordinary contrivances of laws and initial conditions in our universe. These contrivances are a priori more probable under the presupposition of a theistic worldview than under the assumption of a pointless universe. As we saw, however, *theistic design* is only one interpretation among others (interpretation 1). One could mention the ancient Greek notion of the *eternal necessity* of the world (interpretation 2), the chaotic cosmological idea of the *high probability* of cosmos (interpretation 3), and the aforementioned *many world interpretation* (interpretation 4) (Ellis 1993, 372–276).

Note again, however, that these interpretations are not necessarily rivals. In particular, the theistic notion of design is flexible enough to accommodate elements of all three other interpretations. The design idea is compatible with the doctrine of the world's necessity (interpretation 2), provided that God is not the creator of this world but only a formative principle who shapes an already self-existing world. Thus understood, the notion of divine design would be akin to Plato's concept of God as Demiurge (in the *Timaeus*) but would be different from the Jewish, Christian, and Muslim conceptions of God.

With respect to interpretations 3 and 4, God may be invoked as a higher-order explanation for the mathematics of probability or even for the mathematical structures that may allow for different physical laws and different initial conditions in multiple other universes. Thus, one could imagine different types of "meta-anthropic" design in order to account for the contingencies of the orders of the universe.[9]

My aim here is not to review the pros and cons of the anthropic design argument in itself but to clarify its philosophical nature. First, the design argument refers here to strictly *universal* features of the world; divine design is invoked to explain the framework of reality (with its bearings on life and consciousness) without claiming to explain particular features such as the development of elephants, dolphins, and humans. (The term "anthropic" is here misleading since the argument does not concern humanity in particular but life and consciousness in general.)

Second, the meaning of the term "divine designer" is highly ambiguous. The design argument suggests that God is utterly transcendent in relation to the designed world, since a designer is imposing an order on a given material or object. *Who* or *what* the divine is remains unspecified. Usually, the divine designer is tacitly assumed to be the creator of all-that-is (and not only a forming principle), but historically as well as logically the idea of creation is *not* entailed in the notion of design. God could well be

222 PHILOSOPHICAL AND RELIGIOUS PERSPECTIVES

a designer without being the world's creator. In fact, this is the Platonic position revived in the twentieth century by the mathematician-philosopher Alfred North Whitehead. Nor is the biblical notion of God as creator always combined with the idea of a "grand divine master plan." As a matter of fact, the teleological idea of a divine design cannot be inferred from the Old and New Testament texts themselves but is the result of a later Christian appropriation of Platonic and Aristotelian philosophy.[10] The semantic fuzziness of the very notion of design demonstrates that "design" is not an indigenous religious concept. Rather, it is a second-order theoretical construct used by philosopher-theologians to defend the rationality of religious belief. Helpful as the concept of design may be in the context of a philosophical theology, it is not of primary concern in religious life.

Third, the theistic interpretation of the anthropic principle gains its strongest plausibility from the need to explain the *extremely narrow fine-tuning* of the laws and boundary conditions necessary for the development of intelligent life. It is this explanatory need that is satisfied by the reference to a divine designer who deliberately selects the laws and boundary conditions with respect to their intrinsic fruitfulness. A balanced theological position will have to notice both the strengths and weaknesses of the concept of design. Earlier we have seen some of its strengths. One of its weaknesses, however, is that a theology that has so fallen in love with fine-tuning will tend to see an enemy in the *robustness* that is so characteristic of self-organizing systems. But how can theology spell out the significance of self-organizational principles that are not in need of fine-tuning?

Why Design Arguments Are Misplaced in Relation to Self-Organizing Systems

For several reasons, the "laws of complexity" (or rather the general tendencies toward complexification) underlying the emergence and further propagation of complex orders differ from the laws of physics as we usually think of them. These "laws" deal with the cooperative behavior of huge systems that cannot be understood on the basis of self-identical physical particles. Volcanos and ecosystems certainly obey the fundamental laws of physics, yet their behavior cannot be explained by reference to the physicochemical constituents of those systems. The propensities of complex systems exemplify a *synchronous* holistic influence without which one cannot account for the structures that emerge only at the level of the system-as-a-whole.[11] But there is also a *diachronous* aspect to the laws of complexity. Since the systems produce themselves in the course of time and since the rules for self-formation are built into the systems themselves (they don't have the status of Platonic formative principles), the *probability rates are changing during history*. As pointed out by Karl R. Popper, if we want to take probabilities seriously from an ontological point of view, they are "as real as forces" and not a set of mere abstract possibilities. By changing the situation, the probabilities are changing also (Popper 1990, 12–15).

To me this suggests that we should take leave of the older thought model of nature as "the one great chain of being" pouring down as an emanation from a Supreme Being or (in the naturalist translation) as the immediate expression of the basic physical entities. Rather, nature is a continuous "story of becoming" whose outcome can only be predicted in general terms; the important details are open for a historical determination (see Bartholomew 1984). What matters is not only "What is out there in nature?" (on the constitutive level of physics) but also "How does nature work?" (within complex higher-level orders). We have moved away from a materialistic clockwork picture of the universe to an image of the world as consisting of local, very peculiar networks floating in a wider cosmic network governed by physical laws.

Now, which are the theological options in the light of this change from being to becoming? How can theology creatively respond to this new inquiry into the nature of self-organized systems? We should begin by facing how different the features of self-organization are from those features addressed by the anthropic principle.

1. Even if the drive toward pattern-formation can be found at many places in cosmos, the phenomenon of information-based self-organization is not as ubiquitous as gravity. Self-organization is only realized here and there, at once propelled and sheltered by specific environmental conditions. Similarly, in theology, there is a difference between speaking of a *metadesign* that explains the world as a whole (in relation to the anthropic principle) and a *design process* that is related to God's involvement in particular processes within the world.

2. Whereas the notion of God as the external designer is related to God's activity as creator, the idea of God's interaction with a developing world suggests some intimacy (perhaps even a two-way relationship) between God and mundane existence. In the context of the anthropic principle, God is seen as the *context-constitutive* creator of all-that-is (the metadesign). However, if we introduce the concept of design in relation to a self-organizing and self-developing world, we assume that God may guide the particular pathways within the evolutionary phase space.

This notion of a design process presupposes God's activity as a *context-sensitive* interaction with the world of nature.[12]

3. The theory of self-organized complexity needs no divine designer to solve the problem of fine-tuning. This problem is already solved by the flexible, internal dynamics of self-organizing systems. Complexity "can and will emerge 'for free' without any watchmaker tuning the world," according to Per Bak (1997, 48) and Stuart Kauffman (1995, chap. 4).

Provided that this analysis is essentially correct, theology seems to face a stark dilemma. After all, religious life is more interested in the active presence of a providential God in the midst of the world than in a designer God at the edge of the universe. Yet the design argument is viable in the context of the anthropic principle (where its religious significance is rather faint), whereas the design argument seems to be without value in the context of self-organized complexity (where its application would be religiously significant).

Are there ways out of this dilemma? I believe there are at least two options for theology: a *causal* approach in the tradition of philosophical theology and a more qualitative and *descriptive* approach that aims to redescribe the world rather than explaining it. I believe that the latter qualitative approach (which is relatively unconcerned about design) is the more congenial to the spirit of the sciences of complexity and at the same time more appealing from a religious point of view. I intend to show, however, that the two theological options—the first concerned with divine causality and the other concerned with meaning—are indeed supplementary.

The Causal Approach: Reconnecting Design and Self-Organization

The first option consists of an attempt to reconnect the two lines of argumentation that I have so far, for the sake of clarity, differentiated from one another. Ontologically, the evolving laws of self-organization cannot be disconnected from the basic laws of physics, since the former always follow trajectories conditioned by the latter.[13] Any fitness landscape—even if it is open-ended—is always finite, rooted as it is in the specific mathematical properties of chemical systems.[14] All biological adaptations move in the phase space of possibilities constrained by the anthropic principle.

Against this background, we might reconcile the *apparent lack of design* at some levels of reality (e.g., the level of evolutionary selection or of self-organized complexity) with the idea of a divine "metadesign" that affords and even favors complexification in the long run. A solution along this line has been developed both by proponents of a Christian reading of the anthropic principle, such as George Ellis (1993, 375), as well as by proponents of a more general theism, such as Paul Davies. Recently, Davies has thus proposed a modified uniformitarianism: God may be compared with a chess legislator who selects certain rules from the set of all possible rules in order to facilitate a rich and interesting play yet leaves open the particular moves to the players (Davies 1998, 155). After all, some laws are inherently fruitful for the development of life; life, however, is not a step-by-step guided process.

I believe that one can add to this view the possibility of an objectively real divine agency within the world in terms of a "design process."[15] In the science-theology dialogue, several options have been proposed in order to conceptualize God's special actions within the world without assuming a view of God as breaking physical laws.

1. God may continuously influence the world in a bottom-up manner through the quantum processes.[16] Since quantum processes can generate mutations at the genetic level and God, in inscrutable ways, may select between otherwise indeterminate quantum processes (as long as the overall probabilities are not violated), there are potential channels everywhere for God's exercise of a continuous influence on the course of evolution.

2. God may act in a top-down manner by constraining the possibilities on the world as a whole and thereby effectively influence the course of evolution (Peacocke 1993, 191–212).

3. Elsewhere I have proposed a third possibility, which focuses on the changeability of the probability patterns throughout evolution (Gregersen 1998). The point here is that the dice of probabilities are not loaded once and for all but are constantly reloaded in the course of evolution. Given any phase space, the probabilities for future development are constantly changing. With every step some new steps are facilitated, whereas the probability of other possibilities is greatly diminished. Such changeability is always taken for granted by the practicing scientist. In fact, one could argue, with philosopher Nancy Cartwright, that the measurement of such capacities or propensities is what the empirical sciences are really about, whereas the search for mathematically universal laws is more philosophically than scientifically motivated (Cartwright 1989). I would not follow her that far, though. But Cartwright makes the important point that even if there are higher-order laws of nature, lower-level probabilities are constantly re-created in the

process of procedure. This applies especially to *autopoietic systems*, that is, systems that are not only self-organizing but also produce (or at least change) the elements of which the systems are made up. Examples of such autopoietic creativity are the production of lymphocytes in the immune system, the formation of neural pathways in the brain, or the invention of words in human language.

Now the question is: How can we account for the preferred pathways of evolution (and their changing probability rates) if these are *not* fully explainable from the underlying laws of basic physics? How can we explain how nature *works*, if the workings of nature cannot be predicted from the constituent level of the parts or from basic laws of physics? The fact is that we cannot, and I suggest that this unpredictability is given by the role of autonomous agents (Kauffman 2001, 49–80).

Again we have an array of metaphysical possibilities. For what is the *preferential principle* (or the principle of exclusion) that explains why nature follows *this* particular route rather than all other energetically possible routes? Is it (1) a dynamical theism; (2) necessity, or a determinism invoking hidden variables; or (3) chance, or ultimate tychism?[17] Again, I believe that we have both a competition between these views and the possibility of combining them. Evidently, a dynamical and process-oriented theism is opposed to determinism and tychism as long as these latter interpretations are taken as ultimate explanations of reality. However, the notion of divine agency is indeed compatible both with natural causation and with a certain range of chance. In fact, only a dynamical theism (including a design process) is able to embrace the other options. The other way around is not possible. An ultimate determinism would exclude not only an ultimate tychism but also a dynamical theism, and also an ultimate tychism would exclude both theism and determinism. By comparing the three explanatory candidates, theism seems to be the more flexible and open cosmological explanation.

Thus, God may act as a preferential principle guiding the pathways of evolution without determining the individual steps of evolution. God would then not act as a *triggering cause* (like my fingertip on the computer keyboard) but rather as a *structuring cause* that constantly wires and rewires the probability rates of self-organizing processes (like the computer programmer who makes the machine do what I want it to do). In general, what would support a dynamical (nonuniform) theism is a situation where one could say that the overall fruitfulness of self-organizational processes is *more* than one would a priori expect knowing the general laws of physics. There must be more than pure regularity, and more than pure chance, if the notion of God's continuous yet nonuniform activity is to be plausible. A delicately coordinated balance between law and chance seems to be congenial to the hypothesis of a steadfast yet dynamical God who is interacting with a developing world.[18]

The Noncausal Qualitative Approach

I am here already on the way to the qualitative approach, which is also about discerning God in nature but without appealing to specific divine influences on the evolving world. The lead question here is not "What effective difference does God make during evolution?" but rather "How does the nature of the world *express* the nature that God eternally *is*?" Whereas the causal approach understands God as an explanatory principle, the reality of God is here *assumed* (on the basis of religious experience) and subsequently *re-cognized* in the external world. In terminology familiar to theologians, we are here moving from a "natural theology" to a "theology of nature." Self-organizing systems are here seen as prime expressions of God's continuous creativity. Thus the focus is here on how we may theologically *redescribe* the results of the self-organizing processes, which may or may not need special divine guidance.[19]

Earlier I noted how the robustness of self-organizing systems implies a high degree of autonomy, that is, independence from external conditions. I also showed that the theory of self-organization, in this respect, constitutes a new challenge to theology, since self-organization is not in need of any specific fine-tuning. However, complexity theory also offers new options for theology if theology can learn to redescribe, in the language of religion itself, those important features of self-organizing systems that are open to a religious interpretation. What is awe-inspiring about self-organization is not the delicacy of the coordination of the many parameters (as in the anthropic principle) but the sheer fact that the most variegated processes of diversification and creativity are driven by relatively simple laws. These laws, given time and circumstance, *guarantee* that the world of creation attains increasingly complex levels of orders—"for free."

I therefore suggest that the engagement of theology with complexity studies is not best served by rehabilitating or revising ideas of design. What is of more interest to theology is the possibility that a *principle of grace* seems to reign in the creative orderliness of nature. Redescribed in a Judeo-Christian perspective, the theory of self-organization suggests that God is not a remote, acosmic designer of a world but that God is the *blessing God* who creates by bestowing on nature a capacity for fruitful, albeit risky, self-development. The transcendence of God—revealed in the richness of possible pattern formations—is at work *within* the world. As such, God is not identical with specific evolutionary patterns, but God is viewed as the wellspring of the always unprecedented and always changing configurations of order. *God creates the world by giving free freedom—gratuitously. God creates by creating creativity.*

Accordingly, we might need to redescribe our traditional expectations concerning the orderliness of the world. The lesson to be learned both from evolutionary theory and from self-organized criticality is that the laws of nature (as selected by God and instantiated in our world) necessarily include disturbances along the road (Gregersen 2002). Crises evoked by disturbances are necessary means to produce variability and thus enable new high-level configurations of order. Complexity theory may thus prompt theology to rethink its inherited idea of teleology in a manner that appreciates the long-term positive effects of avalanches. States of "harmonious" equilibrium are recurrently punctuated by "catastrophic" avalanches—and both equilibrium *and* catastrophes are governed by the same underlying principles of complexity. The instability of complex systems seems to be the inevitable price to be paid for the immense creativity in evolution. As pointed out by Per Bak in the last sentence of *How Nature Works*, "The self-organized critical state with all its fluctuations is not the best possible state, but it is the best state that is dynamically achievable" (Bak 1997, 198).

Redescribed theologically, evolution is not a risk-free thing but includes the labor of nature. The groaning and suffering of creation is thus not to be seen as specifically designed by a malicious, all-determining God. Setbacks and suffering are part of the package deal of participating in a yet-unfinished creation. However, there is also a principle of grace built into the whole process (Gregersen 2001b, 197–201). There is, taken on a whole, a surplus of life-intensity promised to those who take up the chances of collaboration.

Life is not only a fight for survival, a zero-sum game where one part only wins if the other part loses. As pointed out by Robert Wright in his book *Nonzero*, there is a "non-zero-sumness" built into the structure of the world. "In non-zero-sum games, one player's gain needn't be bad news for the other(s). Indeed, in highly non-zero-sum games the players' interests overlap entirely" (Wright 2000, 5). For complexity is the evolution of coevolution, in which cooperation plays an inextricable role. In Wright's eloquent words,

> biological evolution, like cultural evolution, can be viewed as the on-
> going elaboration of non-zero-sum dynamics. From alpha to omega,
> from the first primordial chromosome up to the first human beings,
> natural selection has smiled on the expansion of non-zero-sumness.
> (Wright 2000, 252)

It is not difficult to redescribe in religious terms this world of non-zero-sumness. Both a demand and a promise is built into the fabric of reality: Only those who are prepared to lose their life will gain it; eventually, they will not gain it over against others but in company with others. In this roundabout way, the world can be said to be designed for self-organization, including both the demand of giving up established pathways of evolution and the promise of becoming more than we would have been without going through the travail of risky self-organization.

After all, at least a Christian theology should be familiar with the Gospel of John's view of the divine Word or Logos that penetrates all that is and that remains the generative Pattern of all patterns of life (John 1:3–4). It is this universal Pattern that is assumed by Christians to have made manifest its everlasting urge in the life story of Jesus Christ. Christians see the historical figure of Jesus as the prime parable of God exactly because he gave up his own existence in order to produce further life. For "unless a grain of wheat falls into the earth and dies, it remains just a single grain; but if it dies, it yields much" (John 12:24).

On this view, it is to be expected that the creation of order includes the experience of setbacks and death as part of God's way of giving birth to new creations. Self-sacrifice as well as self-productivity are built into the structure of the world: a world designed for self-organization.

NOTES

This chapter owes much to conversations with the other symposium participants, especially to points made by Paul Davies, William Dembski, Charles Harper, Harold Morowitz, Arthur Peacocke, and Stuart Kauffman. Thanks also to Kees van Kooten Niekerk, Robert John Russell, Peter Scott, and Mogens Wegener for their thoroughgoing critique of the chapter, and to the Complexity Group of the Danish Science-Religion Forum, which was prepared to devote an evening to discussion of its theses.

1. The tale is told, among others, by Waldrop (1992) and Lewin (1993). A broader (and less Americano-centric view) is presented in Coveney and Highfield (1995).

2. This is underlined in the recent collection of essays edited by André Brack (1998). The editor points to the fact that most of the Earth's early geo-

logical history is erased by later events, and that current theories in the field, many of which are untested, have a puzzle-like character.

3. A recent attempt to pursue the Thomistic thought model (moderated through Bernard Lonergan) is Happel (1995). Happel points out some important implications of this view, namely, that God never acts "directly" but always "mediated" in the world of nature and that "there is no other causal nexus than the self-organization of the entity itself" (197). Thus, this teleological-instrumental model is friendly to a general naturalist perspective: "Rocks cooperate [with God] as rocks, plants as plants, and dogs as dogs" (198). However, it is exactly this sweeping character of the teleological thought scheme that limits its explanatory power. Explaining everything does not, after all, explain the distinctive features of our natural world.

4. I am here following the definition by Barrow and Tipler in their landmark book *The Anthropic Cosmological Principle* ([1986] 1996, 21).

5. See Deutsch [1997] 1998, 32–53, on the double split-experiment, and 275–285, on the multiverse.

6. Some (but not all) of these unique capacities are also pointed out by Barrow and Tipler ([1986] 1996, 524–548).

7. In fact, the substance of Denton's argument of carbon's unique properties for life also questions the hope that carbon-based information in a far future could be transferred into a silicon medium and thus make life everlasting. See Frank Tipler's argument that the persistence of intelligent life is only possible if "an exact replica of ourselves is being simulated in the computer's minds of the far future" (Tipler 1994, 227).

8. See the concise overview in Leslie ([1989] 1996, 25–56), probably the best book on the anthropic principle.

9. See the original proposal in Russell (1989, 196–204).

10. In English Old Testament translations, the reference to what God has "purposed" (e.g., Jer. 4:28) or "planned" (e.g., Isa. 14:24–27; 19:12) is usually derived from the Hebrew verbs *zamas* (meaning "thinking and doing something deliberately") and *azah* (meaning "forming something," like a pot, and "counseling, willing, and making real what one wills"), see Jenni and Westermann (1997, 566–568). When reading English translations of the Bible, one should always bear in mind that Indo-European translations of *zimmah* as "purpose" and *ezah* as "plan/counsel" always refer to substantivized Hebrew verbs that connote a circumstantial, yet very intensive, divine intention-and-action (e.g,. Isa. 46:9). God pursues this or that purpose, but it is nowhere assumed that God has a fixed plan for the history of creation in general. In the New Testament, by contrast, we do find some references to God's one purpose for all-that-is (God's will, *boulé*, God's predetermination, *prothesis*, or God's plan, *oikonomia*) but the purpose referred to here is always the purpose of salvation, not a general cosmic plot (e.g., Eph. 1:4–11). Thus, in the Old Testament, we have a highly context-bound concept of design: a theology of God's will in relation to historical situation (a local teleology, if one so wishes); in the New Testament we have a concept of global design, but a design about

eternal salvation (a transcendent global teleology), never about a mundane harmony (an immanent global teleology), see Gregersen (2000). Historically speaking, the idea of a global this-worldly design is derived from late Jewish Apocalyptic (texts that did not find entrance into the canonical writings) in combination with the great Platonic idea of an artistic God (*deus artifex*). This teleological idea was appropriated in the early Middle Ages by Platonizing Christians and became later, in early modernity, a standard concept of natural theology. Against this background, it is somewhat paradoxical that conservative evangelicals today often appeal to the Enlightenment concept of the designer God as an identity mark of "orthodox" Christianity.

11. On different versions of whole–part causation, see Peacocke 1999.

12. On this difference, see Gregersen 1997, 177–181.

13. See, in this volume, the contributions of Davies, Stewart, Loewenstein, and Morowitz.

14. Stewart 1998. The importance of the finitude of phase space was made clear to me by George Ellis (personal communication, Cape Town, August 8–9, 1999).

15. Here, as elsewhere in this chapter, "process" does not allude to the specifics of Whiteheadian "process theology."

16. See many voices in the Vatican/CTNS project on "Scientific Perspectives on Divine Action," including Robert J. Russell, Nancey Murphy, and George Ellis; see Russell et al. 1998.

17. Note that the option of "infinite possibilities" (comparable to the many world hypothesis) is not viable here, since the phase space of our universe is finite.

18. Note here similarities and dissimilarities to the intelligent design movement. The penultimate sentence may seem to resonate with William Dembski's "explanatory filter": Intelligent design can be inferred where features *cannot* be explained by law or by chance (1998, 36–66). The last sentence, however, makes the difference: It is exactly the delicate (natural!) interplay between law and contingency that is open to a religious interpretation.

19. On the difference between (causal) explanation and (semantic) redescription, see Gregersen 1994, 125–129, further elaborated in Gregersen 2001a. See also van Huyssteen 1998, 125–128.

REFERENCES

Bak, Per. 1997. *How Nature Works: The Theory of Self-Organized Criticality*. New York: Oxford University Press.

Barrow, John B. [1991] 1992. *Theories of Everything: The Quest for Ultimate Explanation*. New York: Ballantine Books.

Barrow, John D., and Frank J. Tipler. [1986] 1996. *The Anthropic Cosmological Principles*. Oxford: Oxford University Press.

Bartholomew, D. J. 1984. *God of Chance*. London: SCM Press.

Brack, André, ed. 1998. *The Molecular Origins of Life: Assembling Pieces of the Puzzle*. Cambridge: Cambridge University Press.

232 PHILOSOPHICAL AND RELIGIOUS PERSPECTIVES

Carter, Brandon. [1974] 1998. "Large Number Coincidences and the Anthropic Principle in Cosmology." In *Modern Cosmology and Philosophy*, edited by John Leslie. Amherst, Mass.: Prometheus Books.

Cartwright, Nancy. 1989. *Nature's Capacities and Their Measurement*. Oxford: Clarendon Press.

Cilliers, Paul. 1998. *Complexity and Postmodernism: Understanding Complex Systems*. London: Routledge.

Davies, Paul. 1998. "Teleology without Teleology: Purpose through Emergent Complexity." In *Evolutionary and Molecular Biology: Scientific Perspectives on Divine Action*, edited by Robert J. Russell et al. Vatican City: Vatican Observatory Publications/Berkeley: CTNS.

de Duve, Christian. 1998. "Clues from Present-day Biology: The Thioester World." In *The Molecular Origins of Life: Assembling Pieces of the Puzzle*, edited by André Back. Cambridge: Cambridge University Press.

Dembski, William A. 1998. *The Design Inference: Eliminating Chance through Small Probabilities*. Cambridge: Cambridge University Press.

Dembski, William A. 1999. *Intelligent Design: The Bridge between Science and Theology*. Downers Grove, Ill.: InterVarsity Press.

Denbigh, K. G. 1975. *An Inventive Universe*. London: Hutchinson.

Denton, Michael J. 1998. *Nature's Destiny: How the Laws of Biology Reveal Purpose in the Universe*. New York: Free Press.

Deutsch, David. [1997] 1998. *The Fabric of Reality*. London: Penguin.

Dyson, Freeman J. 1997. "The Two Windows." In *How Large Is God? The Voices of Scientists and Theologians*, edited by John Marks Templeton. Philadelphia: Templeton Foundation Press.

Ellis, George F. R. 1993. "The Theology of the Antropic Principle." In *Quantum Cosmology and the Laws of Nature: Scientific Perspectives on Divine Action*, edited by Robert J. Russell, Nancey Murphy, and C. J. Isham. Vatican City: Vatican Observatory Publications/Berkeley: CTNS.

Gombridge, R. F. 1975. "Ancient Indian Cosmology." In *Ancient Cosmologies*, edited by Carmen Blacker and Michael Loewe. London: Allen and Unwin.

Gregersen, Niels Henrik. 1994. "Theology in a Neo-Darwinian World." *Studia Theologica* 48.2: 125–149.

Gregersen, Niels Henrik. 1997. "Three Types of Indeterminacy: On the Difference between God's Action as Creator and as Providence." In *Studies in Science and Theology*. Vol. 3. Geneva: Labor et Fides.

Gregersen, Niels Henrik. 1998. "The Idea of Creation and the Theory of Autopoietic Processes." *Zygon* 33,3: 333–367.

Gregersen, Niels Henrik. 1999. "Autopoiesis: Less than Self-Constitution, More than Self-Organization: Reply to Gilkey, McClelland and Deltete, and Brun." *Zygon* 34,1: 117–138.

Gregersen, Niels Henrik. 2000. "Providence." In *The Oxford Companion to Christian Thought*, edited by Adrian Hastings. Oxford: Oxford University Press.

Gregersen, Niels Henrik. 2001a. "The Creation of Creativity and the Flour-
    ishing of Creation." *Currents in Theology and Mission* 28: 3–4, 400–410.
Gregersen, Niels Henrik. 2001b. "The Cross of Christ in an Evolutionary
    World." *Dialog: A Journal of Theology* 38: 3 (Fall), 192–207.
Gregersen, Niels Henrik. 2002. "Beyond the Balance: Theology in a World
    of Autopoietic Systems." In *Design and Disorder: Scientific and theological
    Perspectives*, 53–91. Edited by Niels Henrik Gregersen, Willem B. Drees,
    and Ulf Görman. Edinburgh: T. and T. Clark.
Griffin, David Ray. 2000. *Religion and Scientific Naturalism: Overcoming the
    Conflicts*. Albany: State University of New York Press.
Happel, Stephen. 1995. "Divine Providence and Instrumentality: Metaphors
    for Time in Self-Organizing Systems and Divine Action." In *Chaos and
    Complexity: Scientific Perspectives on Divine Action*, edited by Robert J. Russell
    et al. Vatican City: Vatican Observatory. Publications/Berkeley: CTNS.
Highfield, Roger. 1995. *Frontiers of Complexity: The Search for Order in a Cha-
    otic World*. London: Faber and Faber.
van Huyssteen, J. Wentzel. 1998. *Duet and Duel? Theology and Science in a
    Postmodern World*. London: SCM Press.
Jacobs, Louis. 1975. "Jewish Cosmologies." In *Ancient Cosmologies*, edited by
    Carmen Blacker and Michael Loewe. London: Allen and Unwin.
Jenni, Ernst, and Claus Westermann. 1997. *Theological Lexicon of the Old Tes-
    tament*. Vols. 1–2. Peabody: Hendrickson.
Kauffman, Stuart. 1995. *At Home in the Universe: The Search for the Laws of
    Complexity*. New York: Oxford Univeristy Press.
Kauffman, Stuart. 2001. *Investigations*. New York: Oxford University Press.
Leslie, John. [1989] 1996. *Universes*. London: Routledge.
Lewin, Roger. 1993. *Complexity: Life at the Edge of Chaos*. London: Phoenix.
Lovejoy, Arthur O. [1936] 1964. *The Great Chain of Being: A Study of the History
    of an Idea*. Cambridge, Mass.: Harvard University Press.
Meyer, Stephen C. 1998. "The Explantory Power of Design: DNA and the
    Origin of Information." In *Mere Creation. Science, Faith and Intelligent
    Design*, edited by William A. Dembski. Downers Grove, Ill.: InterVarsity
    Press.
Peacocke, Arthur. 1993. *Theology for a Scientific Age*. Enl. ed. London: SCM
    Press.
Peacocke, Arthur. 1999. "The Sound of Sheer Silence: How Does God
    Communicate with Humanity?" In *Neuroscience and the Human Person*,
    edited by Robert J. Russell et al. Vatican City: Vatican Observatory/
    Berkeley: CTNS.
Peacocke, Arthur. 2001. *Paths from Science towards God: The End of All Explor-
    ing*. Oxford: Oneworld.
Popper, Karl R. 1990. *A World of Propensities*. Bristol, England: Thoemmes.
Russell, Robert John. 1989. "Cosmology, Creation, and Contingency." In
    Cosmos as Creation: Theology and Science in Consonance, edited by Ted
    Peters. Nashville, Tenn.: Abingdon Press.

Russell, Robert John, William R. Stoeger, S. F. and Francisco F. Ayala, eds. 1998. *Evolutionary and Molecular Biology: Scientific Perspectives on Divine Action*. Vatican City: Vatican Observatory.

Shanks, Niall, and Karl H. Joplin. 1999. "Redundant Complexity: A Critical Analysis of Intelligent Design in Biochemistry." *Philosophy of Science* 66: 268–282.

Smolin, Lee. 1997. *The Life of the Cosmos*. New York: Oxford University Press.

Smolin, Lee. 2000. "Our Relationship with the Universe." In *Many Worlds: The New Universe, Extraterrestical Life, and the Theological Implications*, edited by Steven J. Dick. Philadelphia: Templeton Foundations Press.

Stewart, Ian. 1998. *Life's Other Secret: The New Mathematics of the Living World*. London: Penguin Books.

Tipler, Frank. 1994. *The Physics of Immortality: Modern Cosmology. God and the Resurrection of the Dead*. New York: Doubleday.

Waldrop, M. Mitchell. 1992. *Complexity: The Emerging Science at the Edge of Order and Chaos*. New York: Simon and Schuster.

Weinberg, Steven. 1999. "A Designer Universe?" *New York Review of Books*, October 21.

Welker, Michael. 1991. "What Is Creation? Re-Reading Genesis 1 and 2." *Theology Today* 48: 56–71.

Wright, Robert. 2000. *Nonzero: The Logic of Human Destiny*. New York: Pantheon Books.

INDEX